The Unified Process Inception Phase

Best Practices in Implementing the UP

Scott W. Ambler and Larry L. Constantine,
Compiling Editors

Masters collection from

CRC Press
Taylor & Francis Group
Boca Raton London New York

CRC Press is an imprint of the
Taylor & Francis Group, an **informa** business

CRC Press
Taylor & Francis Group
6000 Broken Sound Parkway NW, Suite 300
Boca Raton, FL 33487-2742

First issued in hardback 2017

ISBN 13: 978-1-138-41226-2 (hbk)
ISBN 13: 978-1-929629-10-7 (pbk)

Visit the Taylor & Francis Web site at
http://www.taylorandfrancis.com

and the CRC Press Web site at
http://www.crcpress.com

Cover art design: Robert Ward and John Freeman

To my aunts and uncles:
Bruce, Shirley, June, Peter, Viola, and Ross.

— Scott Ambler

Table of Contents

Foreword

There was a time when one or two talented programmers with a vision could create useful, even groundbreaking, software applications on their own. Today, though, most significant applications demand the resources of a team of professionals with diverse skills to create and refine a substantial body of code and supporting documents. The most effective teams thoughtfully apply sensible software development processes that suit the nature of their project, its context and constraints, and the organization's culture.

One of the well-established, contemporary software development processes is the Unified Process from Rational Corporation. You can buy books that describe the Unified Process in great detail. Such books tell you what to do, but they provide scant guidance about how to do it practically, effectively, and efficiently. Editors Scott Ambler and Larry Constantine close the gap in this book and the others in the series. Scott and Larry bring a vast background of software experience, insight, and perspective to the table. In this book, they collect the even broader experience, insight, and perspective contributed by dozens of authors to more than ten years of *Software Development* and *Computer Language* magazine. The forty-one articles, together with Scott and Larry's additional commentary, provide a wealth of guidance on how to tackle the first phase of the enhanced lifecycle for the Unified Process, the Inception phase. Few of these articles describe projects that specifically followed the Unified Process. However, Scott and Larry have selected key papers that address important recurring themes in software development — best practices — and aligned them with the Unified Process phases. Every software engineer and project manager should know of these best practices and understand how to apply them to whatever process their project uses.

The Inception phase is perhaps the most critical of all. It deals with the "fuzzy front end" of the project, in which the team lays the foundation for success — or for failure. If you don't have a good understanding of your customers' functional, usability, and quality needs, it doesn't matter how well you execute the later phases of Elaboration, Construction, Transition, and Production. If you begin with unclear business objectives, ill-understood user requirements, unrealistic plans and schedules, or unappreciated risks, your project is born with one foot already in the grave.

The Unified Process recognizes that key process workflows, including requirements, implementation, test, and project management, span multiple lifecycle phases. This book therefore includes articles that provide specific recommendations on how to perform the critical activities in each workflow during the Inception phase. Scott Ambler and Larry Constantine also recognize that no single process or methodology can supply a formulaic solution to the diverse challenges and situations a software development organization will encounter. Nonetheless, this book will go a long way toward helping you apply the Unified Process to your projects. And even if you don't care about the Unified Process, the articles provide a solid foundation of current thinking on software engineering best practices.

Karl E. Wiegers
Principal Consultant at Process Impact

Preface

A wealth of knowledge has been published on how to be successful at developing software in *Software Development* magazine and in its original incarnation, *Computer Language*. The people who have written for the magazine include many of the industry's best known experts: Karl Wiegers, Steve McConnell, Ellen Gottesdiener, Jim Highsmith, Warren Keuffel, and Lucy Lockwood, to name a few. In short, the leading minds of the information industry have shared their wisdom with us over the years in the pages of this venerable magazine.

Lately, there has been an increased focus on improving the software process within most organizations. This is in part due to the Year 2000 (Y2K) debacle, to the significant failure rate of large-scale software projects, and to the growing realization that following a mature software process is a key determinant in the success of a software project. In the mid-1990s, Rational Corporation began acquiring and merging with other tool companies. As they consolidated the companies, they consolidated the processes supported by the tools of the merging companies. The objective of the consolidation was to arrive at a single development approach. They named the new approach the Unified Process. Is it possible to automate the entire software process? Does Rational have a complete toolset even if it is? We're not so sure. Luckily, other people were defining software processes too, so we have alternate views of how things should work. This includes the OPEN Consortium's OPEN process, the process patterns of the Object-Oriented Software Process (OOSP), and Extreme Programming (XP). These alternate views can be used to drive a more robust view of the Unified Process, resulting in an enhanced lifecycle that more accurately reflects the real-world needs of your organization. Believing that the collected wisdom contained in *Software Development* over the years could be used to flesh-out the Unified Process — truly unifying the best practices in our industry — we undertook this book series.

Why is a software process important? Step back for a minute. Pretend you want to have a house built and you ask two contractors for bids. The first one tells you that, using a new housing technology, he can build a house for you in two weeks if he starts first thing tomorrow, and it will cost you only $100,000. This contractor has some top-notch carpenters and plumbers that have used this technology to build a garden shed in the past, and they're willing to work day and night for you to meet this deadline. The second one tells you that she needs to discuss what type of house you would like built, and then, once she's confident that she understands your needs, she'll put together a set of drawings within a week for your review and feedback. This initial phase will cost you $10,000, and, once you decide what you want built, she can then put together a detailed plan and cost schedule for the rest of the work.

Which contractor are you more comfortable with — the one that wants to start building or the one that wants to first understand what needs to be built, model it, plan it, then build it? Obviously, the second contractor has a greater chance of understanding your needs, the first step for successfully delivering a house that meets them. Now assume that you're having software built — something that is several orders of magnitude more complex and typically far more expensive than a house, and assume once again that you have two contractors wanting to take these exact same approaches. Which contractor are you more comfortable with? We hope the answer is still the second one; the one with a sensible process. Unfortunately, practice shows that most of the time, organizations appear to choose the approach of the first contractor; that of hacking. Of course, practice also shows that our industry experiences upwards of 85% failure rate on large-scale, mission-critical systems. (In this case, a failure is defined as a project that has significantly overrun its cost estimates or has been cancelled outright.) Perhaps the two phenomena are related.

In reality, the situation is even worse than this. You're likely trying to build a house and all the contractors that you have available to you only have experience building garden sheds. Even worse, they've only worked in tropical parts of the world and have never had to deal with the implications of frost, yet you live in the backwoods of Canada. Furthermore, the various levels of Canadian government enforce a variety of regulations that the contractors are not familiar with — regulations that are constantly changing. Once again, the haphazard approach of the first contractor is likely to get one into trouble.

The Inception Phase

In the enhanced lifecycle for the Unified Process, the Inception phase is the first of five phases — Inception, Elaboration, Construction, Transition, and Production — that a release of software experiences throughout its complete lifecycle. The primary goal of the Inception phase is to set a firm foundation for your project. To accomplish this, you will need to:

- justify both the system itself and your approach to developing/obtaining the system,
- describe the initial requirements for your system,
- determine the scope of your system,
- identify the people, organizations, and external systems that will interact with your system,
- develop an initial risk assessment, schedule, and estimate for your system, and
- develop an initial tailoring of the Unified Process to meet your exact needs.

When you step back and think about it, the most important thing that you can do is ensure that your system, and your approach to it, is justified (i.e., that you have a business case). If the project doesn't make sense — either from an economic, technical, or operational point of view — then you shouldn't continue. Seven out of eight large-scale projects fail. Without a firm foundation, an architecture that will work, a realistic project plan, and a committed team of professionals, your project is very likely going to be one of the seven failures.

This book collects articles written by industry luminaries that describe best practices in these areas. One goal of this book, and of the entire series, is to provide proven alternative approaches to the techniques encompassed by the Unified Process. Another goal is to fill in some of the gaps in the Unified Process. Because the Unified Process is a development process, not a software process that covers development and the operations and support of software once in production, it inevitably misses or shortchanges some of the concepts that are most important for software professionals. Fortunately, the writers in *Software Development* have taken a much broader view of process scope and have filled in many of these holes for us.

About This Series

This book series comprises four volumes: one for the Inception phase, one for the Elaboration phase, one for the Construction phase, and a fourth one for the Transition and Production phases. Each book stands on its own, but for a complete picture of the entire software process you need the entire series. The articles presented span the complete process without duplication among the volumes.

Overall organization of this book series.

Workflow/Topic	Inception Phase (volume 1)	Elaboration Phase (volume 2)	Construction Phase (volume 3)	Transition and Production Phases (volume 4)
Business Modeling	X	X		
Requirements	X	X		
Analysis and Design		X	X	
Implementation			X	
Test	X	X	X	X
Deployment				X
Operations and Support				X
Configuration and Change Management			X	
Project Management	X	X	X	X
Environment	X			X
Infrastructure Management		X	X	X

It has been a challenge selecting the material for inclusion in this compilation. With such a wealth of material to choose from and only a limited number of pages in which to work, narrowing the choices was not always easy. If time and space would have allowed, each of these books might have been twice as long. In narrowing our selections, we believe the articles that remain are truly the *crème de la crème*. Furthermore, to increase the focus of the material in each book, we have limited the number of workflows that each one covers. Yes, most workflows are pertinent to each phase; but as the previous table indicates, each book covers a subset to provide you with a broader range of material for those workflows.

About the Editors

Scott Ambler

An avid reader of *Computer Language* and then *Software Development* for years, I started writing for the magazine in 1995 and eventually became the object columnist in 1997. I started developing software in the early 1980s, writing code in languages such as Fortran and Basic, and later in the mid-1980s in Turing (don't ask), C, Prolog, and Lisp. In the late 1980s, I realized that there was more to life than programming and started picking up skills in user interface design, data modeling, process modeling, and testing while I programmed in COBOL and a couple of fourth-generation languages for IBM mainframes. Disillusioned with structured/procedural techniques, in 1990, I discovered objects and readily jumped into Smalltalk development, then into C++ development, then back to Smalltalk. Having worked at several organizations in mentoring and architectural roles, I decided to combine that experience and apply my skills gained as a teaching assistant at the University of Toronto and get into professional training in the mid-1990s. I quickly learned several things. First, that although I liked delivering training courses (and still do so today), I didn't want to do it full time. Second, and of greater importance, I learned how to communicate complex concepts in an easy-to-understand manner, such as how to develop object-oriented software. This led to my first two books, *The Object Primer* (Cambridge University Press, 1995), now in its second edition (2000), and *Building Object Applications That Work* (Cambridge University Press, 1997/1998), which describe the fundamentals of object technology from a developer's point of view. I then decided to follow up with two books that describe the Object-Oriented Software Process (OOSP) in *Process Patterns* (Cambridge University Press, 1998) and *More Process Patterns* (Cambridge University Press, 1999), focusing on the hard-won experiences that I gained working for one of Canada's leading object technology consulting firms. Since then I've helped several organizations, large and small, new and established, in a variety of industries to improve their internal software processes. My latest writing endeavors include this book series as well as co-authoring *The Elements of Java Style* (Cambridge University Press, 2000). I am now President of Ronin International (www.ronin-intl.com), a Denver-based process and software architecture consulting firm, and a freelance writer with my own web site, www.ambysoft.com, where I post a variety of white papers. I think I've found my niche.

Larry Constantine

My association with *Software Development* and its forerunner, *Computer Language*, has been both long and fruitful, and my association with software development and computer language goes back even further. From my first Fortran program back in the dark ages of computing, I have been keenly interested in figuring out how to do things better and to help others do them better — interests that soon led me beyond technology into management and process issues, as well as the essential matter of the usability of the products we design and build. Throughout my nearly 40 years in the field, I have continued to criss-cross that river that too often divides the people side from the technology side. In my view, success in software development hinges on an understanding and a mastery of material from both sides of this divide, and this has been reflected in my writing for the magazine and elsewhere. That work now spans over 150 articles and papers and 14 books, including, now, this collaborative compilation with Scott Ambler. With Scott's concurrence, some of my own columns and articles in the magazine have been included in these volumes. Others appear in *The Peopleware Papers* (Prentice Hall, 2000), which reprints in its entirety the contents of my long-running "Peopleware" column, and in *Managing Chaos: The Expert Edge in Software Development* (Addison-Wesley, 2000), which incorporates the best from the popular Software Development Management Forum that appears at the back of every issue. In recent years, my professional interests have been particularly focused on increasing the usability of software, which has led to the development of usage-centered design and to the book with Lucy Lockwood, *Software for Use: A Practical Guide to the Models and Methods of Usage-Centered Design* (Addison-Wesley, 1999). The magazine honored us by giving that book the Jolt Product Excellence Award for best book of 1999. Of late, it seems I cross oceans even more often than rivers, because, although I live in the United States, I also teach at the University of Technology, Sydney, Australia, where I am an Adjunct Professor of Computing Sciences. Despite the title, I teach a mix of management and design topics. I am also a working trainer, designer, and consultant helping clients around the world build software that is easier to use. With Lucy Lockwood, I founded Constantine & Lockwood, Ltd. (www.forUse.com), where I am Director of Research and Development and am currently working on the integration of usage-centered design with the Unified Process and Unified Modeling Language, among other things.

Acknowledgments

We'd like to thank the following people for their insightful comments during the development of this book: Susan Ambler, John Nalbone, Mark Peterson, Neil Pitman, Doug Smith, Art Staden, and Robert J. White Jr.

Chapter 1

Introduction

What is a software process? A software process is a set of project phases, stages, methods, techniques, and practices that people employ to develop and maintain software and its associated artifacts (plans, documents, models, code, test cases, manuals, etc.). Not only do you need a software process, you need one that is proven to work in practice — a software process tailored to meet your exact needs.

Why do you need a software process? An effective software process will enable your organization to increase its productivity when developing software. First, by understanding the fundamentals of how software is developed, you can make intelligent decisions, such as knowing to stay away from SnakeOil v2.0 — the wonder tool that claims to automate fundamental portions of the software process. Yes, tools are important, but they must support and accommodate the chosen process, and they should not introduce too much overhead. Second, it enables you to standardize your efforts, promoting reuse, repeatability, and consistency between project teams. Third, it provides an opportunity for you to introduce industry best practices such as code inspections, configuration management, change control, and architectural modeling to your software organization. Fourth, a defined software process establishes a baseline approach for greater consistency and future enhancements.

An effective software process will also improve your organization's maintenance and support efforts — also referred to as production efforts — in several ways. First, it should define how to manage change and appropriately allocate maintenance changes to future releases of your software, streamlining your change process. Second, it should define both how to transition software into operations and support smoothly and how the operations and support efforts are actually performed. Without effective operations and support processes, your software will quickly become shelfware.

An effective software process considers the needs of both development and production.

Why adopt an existing software process, or improve your existing process using new techniques? The reality is that software is growing more and more complex, and without an effective way to develop and maintain that software, the chance of achieving the required levels of quality decreases. Not only is software getting more complex, you're also being asked to create more software simultaneously. Most organizations have several software projects in development at one time and have many moare than that in production — projects that need to be managed effectively. Furthermore, the nature of the software that we're building is also changing — from the simple batch systems of the 1970s for which structured techniques were geared toward, to the interactive, international, user-friendly, 7/24, high-transaction, high-availability online systems that object-oriented and component-based techniques are aimed. And while you're doing that, you're asked to increase the quality of the systems that you're delivering, and to reuse as much as possible so that you can work faster for less money. A tall order — one that is nearly impossible to fill if you can't organize and manage your staff effectively. A software process provides the basis to do just that.

Software is becoming more complex, not less.

1.1 The Unified Process

The Unified Process is the latest endeavor of Rational Corporation (Kruchten, 2000), the same people who introduced what has become the industry-standard modeling notation, the Unified Modeling Language (UML). The heart of the Unified Process is the Objectory Process, one of several products and services that Rational acquired when they merged with Ivar Jacobson's Objectory organization several years ago. Rational enhanced Objectory with their own processes (and those of other tool companies that they have either purchased or partnered with) to form the initial version (5.0) of the Unified Process officially released in December of 1998.

Figure 1.1 presents the initial lifecycle of the Unified Process comprised of four serial phases and nine core workflows. Along the bottom of the diagram, you can see that any given development cycle through the Unified Process should be organized into iterations. The basic concept is that your team works through appropriate workflows in an iterative manner so that at the end of each iteration, you produce an internal executable that can be worked with by your user community. This reduces the risk of your project by improving communication between you and your customers. Another risk-reduction technique built into the Unified Process is the concept that you should make a "go/no-go" decision at the end of each phase — if a project is going to fail, then you want to stop it as early as possible. Granted, the important decision points are actually at the end of the Inception and Elaboration phases (by the time you've hit the end of the Construction phase, it's usually too late to cancel). This is an important concept in an industry with upward of an 80%–90% failure rate on large-scale, mission-critical projects (Jones, 1996).

Figure 1.1   The initial lifecycle of the Unified Process.

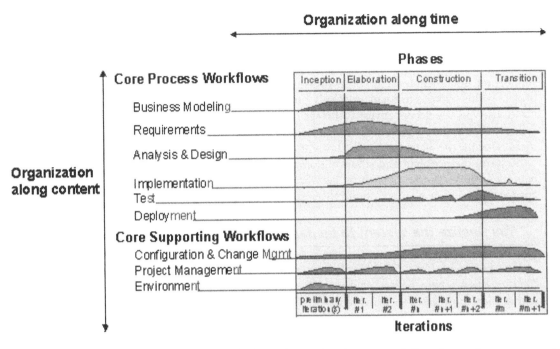

The Inception phase, the topic of this volume, is where you define the project scope and the business case for the system. The initial use cases for your software are identified and the key ones are described briefly. Use cases are the industry standard technique for defining the functional requirements for systems. They provide significant productivity improvements over traditional requirement documents because they focus on what adds value to users as opposed to product features. Basic project management documents are started during the Inception phase, including the initial risk assessment, the estimate, and the project schedule. As you would expect, key tasks during this phase include business modeling and requirements engineering, as well as the initial definition of your environment, including tool selection and process tailoring.

You define the project scope and the business case during
the Inception phase.

The Elaboration phase focuses on detailed analysis of the problem domain and the definition of an architectural foundation for your project. Because use cases aren't sufficient for defining all requirements, a deliverable called a *supplementary specification* is defined which describes all non-functional requirements for your system. A detailed project plan for the Construction Phase is also developed during this phase based on the initial management documents started in the Inception phase.

You define the architectural foundation for your system during the Elaboration phase.

The Construction phase is where the detailed design for your application is developed as well as the corresponding source code. The goal of this phase is to produce the software and supporting documentation to be transitioned to your user base. A common mistake that project teams make is to focus primarily on this phase, often to their detriment because organizations typically do not invest sufficient resources in the previous two phases and therefore lack the foundation from which to successfully develop software that meets the needs of their users. During the Inception and Elaboration phases, you invest the resources necessary to understand the problem and solution domains. During the Construction phase, there should be very little "surprises," such as significantly changed requirements or new architectural approaches — your goal is to build the system.

You finalize the system to be deployed during the Construction phase.

The purpose of the Transition phase is to deliver the system to your user community. There is often a beta release of the software to your users — typically called a pilot release within most businesses — in which a small group of users work with the system before it is released to the general community. Major defects are identified and potentially acted upon during this phase. Finally, an assessment is made regarding the success of your efforts to determine whether another development cycle/increment is needed to further enhance the system. It is during this time that your non-functional requirements, including technical constraints such as performance considerations, become paramount. You will focus on activities such as load testing, installation testing, and system testing — all activities that validate whether or not your system fulfills its non-functional requirements. As you will see in Chapter 3, there is far more to developing requirements than simply writing use cases.

You deliver the system during the Transition phase.

The Unified Process has several strengths. First, it is based on sound software engineering principles such as taking an iterative, requirements-driven, and architecture-based approach to development in which software is released incrementally. Second, it provides several mechanisms, such as a working prototype at the end of each iteration and the "go/no-go" decision point at the end of each phase, which provides management visibility into the development process. Third, Rational has made, and continues to make, a significant investment in their Rational Unified Process (RUP) product (http://www.rational.com/products/rup), an HTML-based description of the Unified Process that your organization can tailor to meet its exact needs. In fact, the reality is that you must tailor it to meet *your* needs because at 3,000+ HTML pages, it comprises far more activities than any one project, or organization, requires. Pick and choose from the RUP the activities that apply, then enhance them with the best practices described in this book series and other sources to tailor a process that will be effective for your team. Accepting the RUP right out of the box is naïve at best — at worst, it is very likely a path to failure.

Attempting to use the Unified Process out of the box is
a recipe for woe and strife. — Neil Pitman

The Unified Process also suffers from several weaknesses. First, it is only a development process. The initial version of the Unified Process does not cover the entire software process. As you can see in Figure 1.1, it is obviously missing the concept of operating and supporting your software once it has been released into production. Second, the Unified Process does not explicitly support multi-project infrastructure development efforts such as organization/enterprise-wide architectural modeling, missing opportunities for large-scale reuse within your organization. Third, the iterative nature of the lifecycle is foreign to many experienced developers, making acceptance of it more difficult, and the rendering of the lifecycle in Figure 1.1 certainly does not help this issue.

The software industry has a capacity for almost infinite self-delusion.
— Capers Jones

In *The Unified Process Elaboration Phase* (Ambler & Constantine, 2000a), the second volume in this series, we show in detail that you can easily enhance the Unified Process to meet the needs of real-world development. We argue that you need to start with the requirements for a process — a good start at which is the Capability Maturity Model (CMM). Second, you should look at the competition — in this case, the OPEN Process (Graham, Henderson-Sellers, and Younessi, 1997; http://www.open.org.au) and the process patterns of the Object-Oriented Software Process (Ambler 1998, Ambler 1999), and see which features you can reuse from those processes. Figure 1.2 depicts the contract-driven lifecycle for the OPEN process and Figure 1.3 depicts the lifecycle of the Object-Oriented Software Process (OOSP), comprised of a collection of process patterns[1]. Finally, you should formulate an enhanced lifecycle based on what you've learned and support that lifecycle with proven best practices.

The Unified Process is a good start, but likely needs to be tailored and
enhanced to meet the specific needs of your organization.

---

1. A *process pattern* is a collection of general techniques, actions, and/or tasks (activities) that solve a specific software process problem taking the relevant forces/factors into account. Just like design patterns describe proven solutions to common software design problems, process patterns present proven solutions to common software process patterns. More information regarding process patterns can be found at the Process Patterns Resource Page, http://www.ambysoft.com/processPatternsPage.html.

Figure 1.2 **The OPEN Contract-Driven lifecycle.**

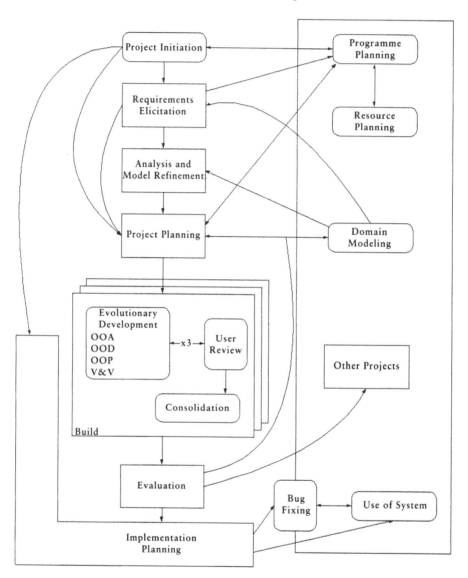

Figure 1.3   The Object-Oriented Software Process (OOSP) lifecycle.

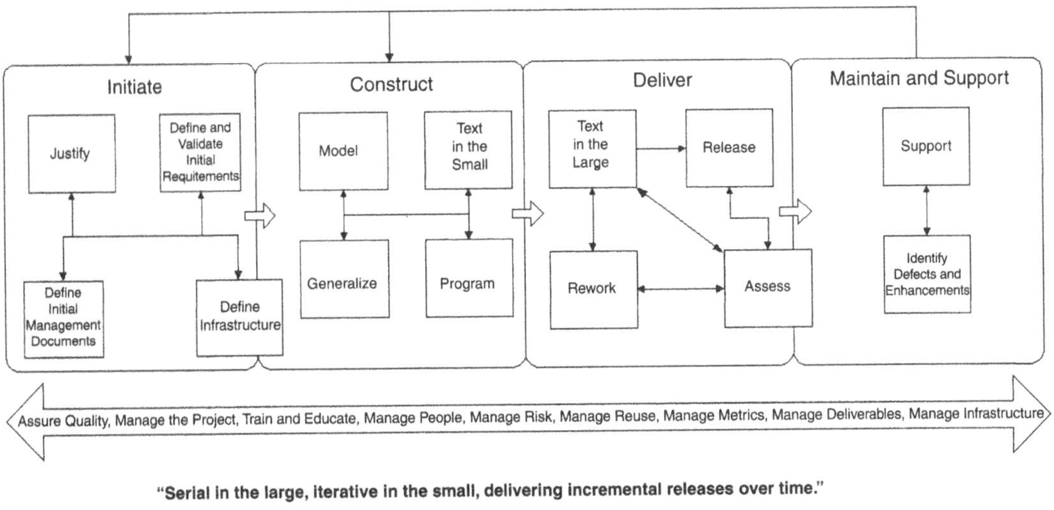

"Serial in the large, iterative in the small, delivering incremental releases over time."

1.2   The Enhanced Lifecycle for the Unified Process

You've seen overviews of the requirements for a mature software process and the two competing visions for a software process. Knowing this, how do you complete the Unified Process? Well, the first thing to do is to redefine the scope of the Unified Process to include the entire software process, not just the development process. This implies that software processes for operations, support, and maintenance efforts need to be added. Second, to be sufficient for today's organizations, the Unified Process also needs to support the management of a portfolio of projects — something the OPEN Process has called "programme management" and the OOSP has called "infrastructure management." These first two steps result in the enhanced version of the lifecycle depicted in Figure 1.4. Finally, the Unified Process needs to be fleshed out with proven best practices; in this case, found in articles published in *Software Development*.

The enhanced lifecycle includes a fifth phase, Production, representing the portion of the software lifecycle after a version of a system has been deployed. Because, on average, software spends 80% of its lifetime in production, the Production phase is a required feature of a realistic software process. Explicitly including a Production phase also enhances the 20% of the lifecycle that is spent in development because it makes it clear to developers that they need to take production issues into account and it provides greater motivation to work towards a common architecture across projects. As the name implies, its purpose is to keep your software in production until it is either replaced with an updated version — from a minor release such as a bug fix to a major new release — or it is retired and removed from production. Note that there are no iterations during this phase (or there is only one iteration depending on how you wish to look at it) because this phase applies to the lifetime of a single release of your software. To develop and deploy a new release of your software, you need to run through the four development phases again.

Figure 1.4   The enhanced lifecycle for the Unified Process.

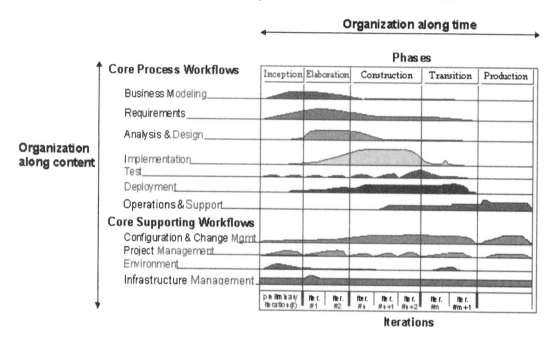

The Production phase encompasses the post-deployment portion of the lifecycle.

Figure 1.4 also shows that there are two new workflows: a core workflow called *Operations & Support* and a supporting workflow called *Infrastructure Management*. The purpose of the Operations & Support workflow is exactly as the name implies: to operate and support your software. Operations and support are both complex endeavors, endeavors that need processes defined for them. This workflow, as well as all the others, span several phases. During the Construction phase, you will need to develop operations and support plans, documents, and training manuals. During the Transition phase, you will continue to develop these artifacts, reworking them based on the results of testing, and you will train your operations and support staff to effectively work with your software. Finally, during the Production phas,e your operations staff will keep your software running, performing necessary backups and batch jobs as needed, and your support staff will interact with your user community in working with your software. This workflow basically encompasses portions of the OOSP's Release stage and Support stage as well as the OPEN Process's *Implementation Planning* and *Use of System* activities. In the Internet economy, where 24/7 operations is the norm, you quickly discover that high quality and high availability is crucial to success — you need an Operations and Support workflow.

The Operations and Support workflow is needed to ensure high quality and high availability of your software.

The Infrastructure Management workflow focuses on the activities required to develop, evolve, and support your organization's infrastructure artifacts such as your organization/enterprise-wide models, your software processes, standards, guidelines, and your reusable artifacts. Your software portfolio management efforts are also performed in this workflow. Infrastructure Management occurs during all phases; the blip during the Elaboration phase represents architectural support efforts to ensure that a project's architecture appropriately reflects your organization's overall architecture. This includes infrastructure modeling activities such as the development of an enterprise requirements/business model, a domain architecture model, and a technical architecture model. These three core models form your infrastructure models that describe your organization's long-term software goals and shared/reusable infrastructure. The processes followed by your Software Engineering Process Group (SEPG) — responsible for supporting and evolving your software processes, standards, and guidelines — are also included in this workflow. Your reuse processes are included as well because practice shows that to be effective, reuse management needs to be a cross-project endeavor. For you to achieve economies of scale developing software, increase the consistency and quality of the software that you develop, and to increase reuse between projects, you need to manage your common infrastructure effectively. You need the Infrastructure Management workflow.

Infrastructure Management supports your cross-project/programme-level activities such as reuse management and organization/enterprise-wide architecture.

If you compare the enhanced lifecycle of Figure 1.4 with the initial lifecycle of Figure 1.1, you will notice that several of the existing workflows have also been updated. First, the Test workflow has been expanded to include activity during the Inception phase. You develop your initial, high-level requirements during this phase — requirements that you can validate using techniques such as walkthroughs, inspections, and scenario testing. Two of the underlying philosophies of the OOSP are that (a) you should test often and early and, (b) that if something is worth developing, then it is worth testing. Therefore, testing should be moved forward in the lifecycle. Also, the Test workflow needs to be enhanced with the techniques of the OOSP's *Test in the Small* and *Test in the Large* stages.

Test early and test often. If it is worth creating, it is worth testing.

The second modification is to the Deployment workflow — extending it into the Inception and Elaboration phases. This modification reflects the fact that deployment, at least of business applications, is a daunting task. Data conversion efforts of legacy data sources are often a project in their own right — a task that requires significant planning, analysis, and work to accomplish. Furthermore, our belief is that deployment modeling should be part of the Deployment workflow — not the Analysis & Design workflow as it is currently — due to the fact that deployment modeling and deployment planning go hand-in-hand. Deployment planning can, and should, start as early as the Inception phase and continue into the Elaboration and Construction phases in parallel with deployment modeling.

> ***Deployment is complex and planning often must start early in development to be successful.***

The Environment workflow has been updated to include the work necessary to define the Production environment — work that would typically occur during the Transition phase. (You could easily do this work earlier if you wish, but the Transition phase often proves the best time for this effort). The existing Environment workflow processes effectively remain the same — the only difference being that they now need to expand their scope from being focused simply on a development environment to also include operations and support environments. Your operations and support staff need their own processes, standards, guidelines, and tools — the same as your developers. Therefore, you may have some tailoring, developing, or purchasing to perform to reflect this need.

The Configuration & Change Management workflow is extended into the new Production phase to include the change control processes needed to assess the impact of a proposed change to your deployed software and to allocate that change to a future release of your system. This change control process is often more formal than what you do during development due to the increased effort required to update and re-release existing software.

> ***Change control management will occur during the Production phase.***

Similarly, the Project Management workflow is also extended into the new Production phase to include the processes needed to manage your software once it has been released. It is light on metrics management activities and subcontractor management, a CMM level 2 key process area, needed by of any organization that outsources portions of its development activities or hires consultants and contractors. People management issues, including training and education as well as career management, are barely covered by the Unified Process because those issues were scoped out of it. There is far more to project management than the technical tasks of creating and evolving project plans. You also need to manage your staff and mediate the interactions between them and other people.

> ***There is far more to project management than planning, estimating, and scheduling.***

1.3   The Goals of the Inception Phase

During the Inception phase, your project team will focus on understanding the initial requirements, determining scope, and organizing the project. To understand the initial requirements, you will likely perform business modeling and essential modeling activities (Constantine and Lockwood, 1999). Essential models are intended to capture the essence of problems through technology-free, idealized, and abstract descriptions. Your resulting design models are more flexible, leaving more options open and accommodating changes more readily in technology. Essential models are more robust than concrete representations, simply because they are more likely to remain valid in the face of both changing requirements and changes in the technology of implementation. Essential models of usage highlight purpose, what it is that users are trying to accomplish, and why they are doing it. In short, essential models are ideal artifacts to capture the requirements for your system.

Your project's scope can be determined through negotiation with project stakeholders as to the applicability of the results of your business and requirements modeling efforts. The optional development of a candidate architecture enables you to determine both the technical feasibility of your project and its scope because you may find that some aspects of your system are better left out of the current release. Significant project management effort also occurs during the Inception phase where you'll organize your future work, including the development of schedules, estimates, plans, and risk assessments.

The focus of the Inception phase is to define and come to agreement with respect to the high-level requirements, vision, and scope of your project, as well as to justify the project and obtain resources to continue work on it. As you work towards these goals, you will create and/or evolve a wide variety of artifacts, such as:

- A vision document
- An initial requirements model (10–20% complete)
- An initial project glossary
- A business case
- An initial domain model (optional)
- An initial business model (optional)
- A development case describing your project's tailored software process (optional)
- An architectural prototype (optional)

The phase is concluded with the Lifecycle Objective (LCO) milestone (Kruchten, 2000). To pass this milestone, you must achieve:

- a consensus between project stakeholders as to the project's scope and resource requirements,
- an initial understanding of the overall, high-level requirements for the system,
- a justification for your system that includes economic, technological, and operational issues,
- a credible, coarse-grained schedule for your entire project,
- a credible, fine-grained schedule for the initial iterations of the Elaboration phase,
- a credible, risk assessment and resource estimate/plan for your project,
- a credible initial tailoring of your software process,
- a comparison of your actual vs. planned expenditures to date for your project, and
- the development of an initial architectural prototype for your system (optional).

A beginning is the time for taking the most delicate care that the balances are correct. — *Maud'Dib*

1.4 How Work Generally Proceeds During the Inception Phase

A fundamental precept of the Unified Process is that work proceeds in an iterative manner throughout the activities of the various workflows. However, at the beginning of each iteration, you will spend more time in requirements-oriented activities and towards the end of each iteration, your focus will be on test-oriented activities. As a result, to make this book easier to follow, the chapters are organized in the general order by which you would proceed through a single iteration of the Inception phase. As Figure 1.4 indicates, the workflows applicable during the Inception phase are:

- Business Modeling (Chapter 2)
- Requirements (Chapter 3)
- Analysis & Design (covered in Vols. 2 and 3, *The Unified Process Elaboration Phase* and *The Unified Process Construction Phase*, respectively)
- Implementation (covered in Vols. 2 and 3, *The Unified Process Elaboration Phase* and *The Unified Process Construction Phase*, respectively)
- Test (Chapter 4)
- Deployment (covered in Vol. 4, *The Unified Process Transition Phase*)
- Operations and Support (covered in Vol. 4, *The Unified Process Transition Phase*
- Configuration & Change Management (covered in Vol. 3, *The Unified Process Construction Phase*)
- Project Management (Chapter 5)
- Environment (Chapter 6)
- Infrastructure Management (covered in Vols. 2 and 3, *The Unified Process Elaboration Phase* and *The Unified Process Construction Phase*, respectively)

1.4.1 The Business Modeling Workflow

The purpose of the Business Modeling workflow, described in detail in Chapter 2, is to model the business context of your system. During the Inception phase, the focus of the Business Modeling workflow is to:

Identify the context of your system.   A context model shows how your system fits into its overall environment. This model will depict the key organizational units that will work with your system, perhaps the marketing and accounting departments, and the external systems that it will interact with. By developing the context model, you come to an understanding of the structure and culture of the organization and the external business environment your system will exist in and support.

Identify the basis for a common understanding of the system and its context with stakeholders.   Your goal is to begin working towards a common understanding of what your project team will deliver to its stakeholders as well as a common understanding of the environment in which you will work. Stakeholders include your direct users, senior management, user management, your project team, your organization's architecture team, and potentially even your operations and support management. Without this common understanding, your project will likely be plagued with politics and infighting and could be cancelled prematurely if senior management loses faith in it.

Model the business.   When you are modeling a business, you want to understand its goals, strengths, weaknesses, and the opportunities and challenges that it faces in the market place. A business model is multifaceted — potentially including a process model, a use-case model, and a conceptual object model. Your project's business model should reflect your enterprise requirements model. Enterprise modeling is a key aspect of the Infrastructure Management workflow which is covered in detail in *The Unified Process Elaboration Phase* (Ambler & Constantine, 2000a) and *The Unified Process Construction Phase* (Ambler & Constantine, 2000b). Business process models show how things get done, as opposed to a use-case model that shows what should be done and can be depicted using UML activity diagrams. Conceptual object models show the major business entities, their responsibilities, and their inter-relationships, and should be depicted using UML class diagrams. Your business model should include a glossary of key terms and important technical terms optional that your project stakeholders need to understand. The business model is important input into your Requirements workflow efforts.

Your business model shows how your system fits into its environment and helps you to evolve a common understanding with your project stakeholders.

1.4.2 The Requirements Workflow

The purpose of the Requirements workflow, the topic of Chapter 3, is to engineer the requirements for your project. During the Inception phase, you will:

Identify the initial requirements.   You need to identify the requirements for your software to provide sufficient information for scoping, estimating, and planning your project. To do this, you often need to identify your requirements to the 10–20% level — to the point where you understand at a high-level what your project needs to deliver, but may not fully understand the details. You are likely to develop several artifacts as part of your requirements model, including, but not limited to, an essential use-case model (Constantine & Lockwood, 1999; Ambler, 2001), an essential user interface prototype (Constantine & Lockwood, 1999; Ambler, 2001), a user-interface flow diagram (Ambler, 2001), and a *supplementary specification* (Kruchten, 2000). A supplementary specification is a "catch-all" artifact where you document business rules, constraints, and non-functional requirements. During the Elaboration phase, you will develop your requirements model to the 80% level and finalize it during the Construction phase.

Use cases are only a small part of your overall requirements model.

Develop a vision for the project.   The vision document summarizes the high-level requirements for a project, including its behavioral requirements, its technical requirements, and the applicable constraints placed on the project.

Elicit stakeholder needs. A software project often has a wide range of stakeholders, including, but not limited to, its end users, senior management, and your operations and support staff. An important part of understanding the requirements of a system is eliciting the needs of its stakeholders so that the overall vision for the project may be negotiated.

Define a common vocabulary. A project glossary should be started during the Inception phase and evolved throughout the entire lifecycle. This glossary should include a consistent and defined collection of business terms applicable to your system. It is also common to include technical terms pertinent to your project — such as *essential modeling, use case, iteration*, and *Java* — that your project stakeholders need to be familiar with.

Define the scope of the project. An important part of requirements engineering is the definition of the boundary, the scope, of your system. Your goal is to define what requirements your project team intends to fulfill and what requirements it will not. For example, if you were developing a banking system, you may decide that the current version will support international transactions, although it will do so in a single currency only — leaving multi-currency support for a future release of the system.

1.4.3 The Analysis and Design Workflow

The purpose of the Analysis and Design workflow is to model your software. During the Inception phase, your modeling efforts are likely to focus on understanding the fundamental business and the requirements of your system — activities of the Business Modeling and Requirements workflows. It is often quite common to identify, and then model, a candidate architecture that is likely to meet your project's needs — an architecture that should be prototyped as part of your Implementation workflow efforts. By showing that your candidate architecture works, you help to show the technical feasibility of your approach, an important aspect of justifying your project to senior management as part of the Project Management workflow.

For projects that involve significant integration with existing systems, you may find that you need to invest resources in understanding the interfaces that you have with those systems — an effort often called *legacy analysis, external interface analysis,* or simply *data analysis.* Your context model, developed as part of your business modeling efforts, will indicate the external systems that yours will interact with. Once these external systems are identified, you may decide to begin work analyzing the interfaces. Ideally, you would merely need to obtain access to the existing documentation, although you will often find that little or no documentation exists and therefore, significant analysis must occur. This work can begin in the Inception phase and continue into the Construction phase. Legacy analysis techniques are described in *The Unified Process Elaboration Phase* (Ambler & Constantine, 2000a).

Analysis of interfaces to legacy systems is often a significant part of the modeling efforts for a project, an activity that often begins during the Inception phase.

1.4.4 The Implementation Workflow

The purpose of the Implementation workflow is to write and initially test your software. During the Inception phase, your implementation efforts will likely focus on the development of a candidate architectural prototype to prove the technical feasibility of your approach. Technical prototyping, also called *proof-of-concept prototyping*, is described in *The Unified Process Elaboration Phase* (Ambler & Constantine, 2000a). You may also decide to develop a user interface prototype, described in Chapter 3, although your team is more likely to focus on the development of an essential user interface prototype early in the project lifecycle, leaving traditional user interface prototyping efforts for the Elaboration and Construction phases.

Prototyping work may occur during the Inception phase.

1.4.5 The Deployment Workflow

The purpose of the Deployment workflow is to ensure the successful deployment of your system. During the Inception phase, the Deployment workflow may include several key activities, such as:

Develop the initial version your deployment plan. Deployment of software, particularly software that replaces or integrates with existing legacy software, is a complex task that needs to be thoroughly planned. Your deployment plan may begin as early as the Inception phase and be evolved throughout the Elaboration and Construction phases.

Develop a deployment model of the current environment. As you work towards an understanding of the context of your system, you may find that you need to model the deployment of the existing systems, at least those applicable to the system that you are building, within the current environment. This model will provide valuable input into the development of a context model — an activity of the Business Modeling workflow — as well as into your legacy analysis efforts that are part of the Analysis and Design workflow. Deployment modeling was originally an activity of the Analysis and Design workflow in the initial version of the Unified Process (Kruchten, 2000), but this work has been moved into the Deployment workflow for the enhanced lifecycle. The reason for this is simple: deployment planning and deployment modeling go hand-in-hand and are the main drivers of your actual deployment efforts. This is a stylistic issue more than anything else.

Work closely with your operations and support departments. You need assistance from your operations and support departments to successfully deploy your software. Your project is more likely to be accepted the earlier you start working with them. You may find that your operations and support departments have stringent quality gates that a new project must pass through in order to be deployed into production — quality gates that must be well understood and planned for.

Deployment of systems can be quite complex. The sooner you begin planning and modeling how your system will be deployed, the better.

1.4.6 The Test Workflow

The purpose of the Test workflow, the topic of Chapter 4, is to verify and validate the quality and correctness of your system. During the Inception phase, you can validate a wide range of project artifacts including your requirements model, your business case, and your project plan through inspections and reviews. A fundamental philosophy of ours is that if you can build it, you can test it, and anything that is not worth testing, is likely not worth building. It is possible to test your requirements model — you can do a user interface walkthrough, a use-case model walkthrough, or even use-case scenario testing (Ambler, 1998b). It is also possible to test your models and project management artifacts — you can perform peer reviews and inspections.

1.4.7 The Configuration and Change Management Workflow

The purpose of the Configuration and Change Management workflow is to ensure the successful deployment of your system. During the Inception phase, this workflow focuses on placing project artifacts under configuration management control. (Because the initial requirements for your system are not accepted until the end of the Inception phase there is nothing yet to place under change control.) Configuration management (CM) is essential to ensure the consistency and quality of the numerous artifacts produced by the people working on a project, helping to avoid confusion amongst team members. Without CM, your team is at risk of overwriting each other's work — potentially losing significant time to replace any lost updates. Furthermore, good CM systems will notify interested parties when a change has been made to an existing artifact, allowing them to obtain an updated copy if needed. Finally, CM allows your team to maintain multiple versions of the same artifact, allowing you to roll-back to previous versions if need be.

Place project artifacts under change control right from the very start of a project.

1.4.8 The Project Management Workflow

The purpose of the Project Management workflow, the topic of Chapter 5, is to ensure the successful management of your project team's efforts. During the Inception phase, the Project Management workflow focuses on several key activities, such as:

Identify potential risks.   The project team must ensure that the project's risks are managed effectively. The best way to do that is to ensure that the riskier and/or more important requirements are assigned to earlier iterations.

Assess the viability of the project.   In addition to showing that a project is economically justifiable, you must also show that it is technologically and operationally feasible as we show with the Justify process pattern (Ambler, 1998b) described in Chapter 5. At the end of the Inception phase — in fact, at the end of every phase — a project viability assessment should be made (also called a "go/no-go" decision) regarding the project.

Develop an initial project plan. You need to develop a coarse-grained project plan — effectively a high-level plan that identifies the major milestones and key deliverables of your team. This plan should also identify the initial iterations that you expect in each phase — iterations identified based on your project's requirements, the topic of Chapter 3. (Each phase, including the Inception phase, is comprised of one or more iterations.) A common mistake that project managers make is to attempt to develop an intricately detailed plan for an entire project at the very beginning, which you will see in Chapter 5 to be an unrealistic goal at best, and a recipe for disaster at worst.

Develop a detailed plan for the Inception phase. Although it is very difficult to develop a detailed plan for the entire project, it is possible to develop one for the immediate time frame — namely the Inception phase. An important part of project management during any project phase is the definition and continual update of a detailed project plan for that phase. Your planning efforts will include the definition of an estimate and a schedule — artifacts that you will update periodically throughout your project.

Navigate the political waters within your organization. An unfortunate reality of software development is that softer issues, such as people management and politics, will affect all projects.

Establish the metrics to be collected during the project. In Chapter 5, you will discover that there is a wide range of metrics available to you to better manage your object-oriented and component-based development efforts. You will also learn that you need to pare down this list to a handful that provide information pertinent to your project, and ideally to your organization as a whole.

Initiate and define relationships with subcontractors and vendors. You may decide to outsource all or a portion of your project to an organization specializing in certain aspects of software development. If so, you will need to identify potential organizations that appear to meet your needs and initiate relationships with them.

Build the project team. Potential team members, including both system professionals and members of the user community, need to be identified and recruited. You generally do not need to recruit everyone on the very first day of your project — not that everyone is likely to be available anyway — but you must be prepared to meet the resource needs of your project throughout its lifecycle in a timely manner. You will find that you need to recruit within your organization for existing staff and perhaps externally as well for "new blood."

Manage the team's efforts. A key activity in project management is to ensure that the team works together effectively and that the personal career goals of each team member are being met to the best of your abilities.

Begin training and education of team members. As you bring people onto your team, you should assess their abilities and career goals, map them to the needs of your project, and obtain training and education to fill any gaps as needed. This training and education effort may include formal courses in specific technologies or techniques, computer-based training (CBT), mentoring, or bag-lunch sessions.

Obtain project funding. A *business case* (Kruchten, 2000) is an artifact that describes the economic justification of your project — important information that can be used to obtain funding for your project. Your business case will also summarize the business context, success criteria, financial forecast, initial risk assessment, and plan for your project.

> *There is far more to project management than planning, estimating, and scheduling.*

1.4.9 Environment Workflow

The purpose of the Environment workflow, the topic of Chapter 6, is to configure the processes, tools, standards, and guidelines to be used by your project team. Most Environment workflow effort occurs during the Inception and Elaboration phases. During this phase, you will perform three important activities:

1. **Tailor your software process.** Although the definition and support of your organization's software process is an activity of the Infrastructure Management workflow, you still need to tailor that process to meet your project's unique needs. In the Unified Process, the tailoring of your software process is called a *development case.*

2. **Standards and guidelines selection/development.** If your organization does not mandate development guidelines, or does not mandate guidelines applicable to the techniques and technologies that you intend to apply on your project, then you will either need to select or develop these guidelines as part of your Environment workflow efforts. The good news is that many development standards and guidelines are already available to your organization. For example, the Unified Modeling Language (UML) (Rumbaugh, Jacobson, Booch, 1999) provides an excellent basis for modeling notation standards and coding guidelines for Java can be downloaded from http://www.ambysoft.com/javaCodingStandards.html. You don't need to reinvent the guidelines wheel — instead, with a little bit of investigation, you often find that you can reuse existing standards and guidelines.

3. **Tool selection and integration.** Software process, architecture, organizational culture, and tools go hand-in-hand. If they are not already imposed upon your team by senior management, you need to select a collection of development tools such as a configuration management system, a modeling tool, and an integrated development environment. Your tools should work together and should reflect the processes that your team will follow.

1.4.10 The Infrastructure Management Workflow

The Infrastructure Management workflow encompasses activities that are outside of the scope of a single project, yet are still vital to your organization's success. During the Inception phase, the Infrastructure Management workflow includes several key activities, such as:

Identify opportunities for reuse. Strategic reuse management is a complex endeavor — one that spans all of the projects within your organization. Your team should strive to: identify and reuse existing artifacts wherever possible, buy instead of build where that makes sense, and produce high-quality artifacts that can potentially be generalized for reuse.

Identify how your project fits into your organization's software portfolio. *Programme Management* (Graham, Henderson-Sellers, and Younessi, 1997) is the act of managing your organization's portfolio of software projects — projects that are either in development, in production, or waiting to be initiated. You should identify where your project fits in the overall picture so that it can be managed accordingly. As you would expect, this is an important part of defining its scope (an activity of the Project Management workflow).

Update your enterprise requirements model. Enterprise requirements modeling (Jacobson, Griss, Jonsson, 1997; Ambler, 1998b) is the act of creating a requirements model that reflects the high-level requirements of your organization. Your project's requirements model should reflect, in great detail, a small portion of this overall model. As you evolve your requirements model, an activity of the Requirements workflow, you will need to work together with the appropriate parties to ensure that the two models are consistent with one another, which may include the update of the enterprise requirements model.

Evaluate the applicability of your organization/enterprise-level architectural models. Although your individual system may have its own unique architecture, it needs to fit in with your organization's overall business/domain and technical architecture (Ambler, 1998b). Your project's architecture should start with the architecture as a base and then ensure that any futuredeviations fit into the overall picture. To accomplish this, you must first begin by understanding the existing architecture currently in place. A major focus of the Elaboration phase (Ambler & Constantine, 2000a) is the definition of your project's architecture.

Perform organization/enterprise-wide process management. Your project team may discover, through its Environment workflow efforts, that existing corporate processes, standards, and guidelines need to be evolved to meet the new needs of the business. If yours is the first project to use iterative, incremental, object-oriented, or component-based development techniques, you may find that your existing software process needs significant rework.

Strategic reuse management, enterprise requirements modeling,
organization/enterprise-wide architecture, and process management
are infrastructure issues that are beyond the scope of a single project.

1.5   The Organization of this Book

This book is organized in a simple manner. There is one chapter for each of the Project Management, Business Modeling, Requirements, Test, and Environment workflows. Each chapter is organized in a straightforward manner — starting with our thoughts about best practices for the workflow, followed by our comments about the *Software Development* articles that we have chosen for the chapter, then the articles themselves. The book finishes with a short chapter summarizing the activities of the Inception phase and providing insight into the Elaboration phase, the next phase of the Unified Process.

The seeds of major software disasters are usually sown in the first
three months of commencing the software project. — Capers Jones

---

2

Chapter 2

Best Practices for the Business Modeling Workflow

Introduction

The purpose of the Business Modeling workflow is to develop an understanding of the business context of your system. Eriksson and Parker (2000), in their book *Business Modeling with UML*, believe that business modeling is important to your development efforts because it acts as the basis for:

- better understanding the key mechanisms of an existing business,
- creating suitable information systems that support the business,
- improving the current business structure and organization,
- showing the structure of an innovated business,
- experimenting with new business concepts or to copy/study the business concepts of a partner or competitor, and
- identifying outsourcing opportunities.

A business model typically contains:

1. A context model that shows how your system fits into its overall environment. This model will depict the key organizational units that will work with your system, perhaps

the marketing and accounting departments, and the external systems it will interact with. By developing the context model, you come to an understanding of the structure and culture of the organization and its external business environment that your system will exist in and support.

2. A high-level use case model, typically a portion of your enterprise model (perhaps modeled in greater detail), that shows what behaviors you intend to support with your system.

3. A glossary of key terms (something that is evolved as part of your Requirements model).

4. An optional, high-level class diagram (often called a Context Object Model, a Domain Object Model, or a Conceptual Object Model) that models the key business entities and the relationships between them.

5. A business process model (traditionally called an analysis data-flow diagram in structured methodologies) depicts the main business processes, the roles/organizations involved in those processes, and the data flow between them. Business process models show how things get done, as opposed to a use-case model that shows what *should* be done.

During the Inception phase, your goal is to begin working towards a common understanding of what your project team will deliver to its stakeholders as well as a common understanding of the environment in which you work. Stakeholders include your direct users, senior management, user management, your project team, your organization's architecture team, and potentially even your operations and support management. Without this common understanding, your project will likely be plagued with politics and infighting and could be cancelled prematurely if senior management loses faith in it.

You need to build consensus among your project's stakeholders.

An important best practice for this workflow is to ensure that your project's business model reflects your enterprise requirements model. Enterprise modeling is a key aspect of the Infrastructure Management workflow which is covered in detail in *The Unified Process Elaboration Phase* (Ambler & Constantine, 2000a) and *The Unified Process Construction Phase* (Ambler & Constantine, 2000b). Figure 2.1 depicts how the business model fits into the scheme of things. In many ways, Figure 2.1 is a context model for your business modeling efforts. The figure shows that the enterprise model drives the development of the business model — effectively, a more detailed view of a portion of your overall business. It isn't purely a top-down relationship, however, because as your business analysts learn more about the business problem that your system addresses, they will share their findings back to the owners of the enterprise model who will update it appropriately. The business model, in turn, drives the development of your requirements model. The Requirements workflow is covered in Chapter 3, which similarly provides feedback for the business model.

Your business model should reflect your enterprise
requirements model (if any).

Figure 2.1　How the business model fits in.

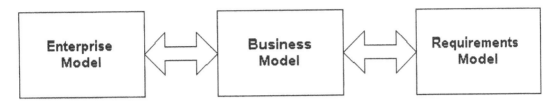

A second best practice with respect to the Business Modeling workflow is to work with a representative range of project stakeholders in order to understand their needs. It isn't enough to simply hear senior management's view of what your system should accomplish, nor is it sufficient to only talk to end users, nor only to their direct managers, nor only to... you get our point. You need to make an effort to understand *everyone's* view of what the system should accomplish and how it should fit into the overall picture.

In the first article in Section 2.1.1, "How the UML Models Fit Together" (March 1998), Scott Ambler takes Figure 2.1 one step further by examining the relationships between the models of the Unified Modeling Language (UML). Modified from his Jolt-award winning book, *Building Object Applications that Work* (Ambler, 1998a), the article describes each UML artifact — such as class diagrams and use case diagrams — and shows how they fit together to form an iterative modeling process. As a business modeler, it is crucial to understand when, why, and how to effectively apply the modeling techniques at your disposal.

A good business modeler understands when, why, and how to effectively apply the models of the UML.

As you will discover in this chapter, another important best practice for you to adopt is the recognition that there is far more to object-oriented and component-based modeling than is supported by the artifacts of the UML. Figure 2.2, taken from *The Object Primer 2nd Edition* (Ambler, 2001), breaks the UML barrier and presents a more realistic picture of how complex your modeling efforts can potentially become. The diagram's boxes, such as Collaboration Diagram and Essential Use Case Model, represent artifacts and the arrows represent "drives" relationships. For example, you see that information in your design class model drives the development of your source code. The artifacts of the UML 1.X are in shaded boxes, putting into stark relief the extent of what is currently missing from the UML. To be sufficient for the needs of real-world development, the UML and the Unified Process need to be enhanced with additional modeling techniques that address the specific needs of your project. Throughout this book series, we present a wide range of modeling techniques — both UML-based and beyond the current scope of the UML — to provide you with a robust selection of modeling options.

The artifacts of the UML are a good start, but they are often not sufficient for large-scale, mission-critical development.

Figure 2.2   The relationships between primary object-development artifacts.

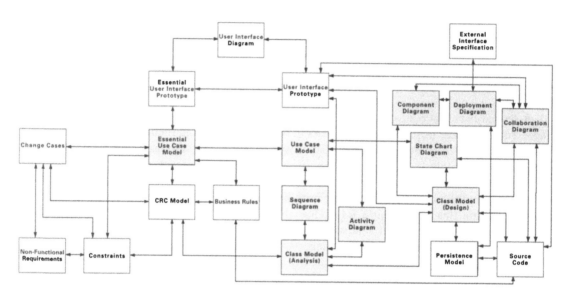

The first step toward creating a usable system is making sure it addresses the customers' needs and that it works the way they expect. In "Data-Based Design" (January 1997, Section Section 2.1.2), Hugh Beyer and Karen Holtzblatt describe an alternative yet complementary approach to business modeling. Let us start by saying that the title of this section doesn't reflect the actual material covered — it has nothing to do with database design. Instead, the authors focus on gathering information (data) about your user community through contextual modeling and then effectively applying that knowledge. By understanding your users — by gathering key data about their needs, their culture, and their work practices — you will be in a significantly better position to build a system that meets their actual needs. Contextual modeling focuses on five artifacts: (1) flow models that show the communication and coordination between people and organizational units, (2) context models that show the cultural and organizational environment, (3) physical models describing the physical environment affecting the work, (4) sequence models showing the ordering of the work over time, and (5) artifact models showing the structure and intent of the artifacts of the work. From the point of view of the Unified Modeling Language (UML), activity diagrams can be used for flow modeling, use-case models for context models, deployment diagrams for physical modeling, and sequence diagrams for sequence modeling. The UML doesn't have a diagram appropriate for artifact modeling at this time, although the free-form nature of this activity leads us to believe that you'd be better off using a simple drawing tool to create these models anyway. Beyer and Holtzblatt present sage advice for successful business modeling — and for business process reengineering in general — which is often a key goal of the Elaboration phase. You will find this article to be an excellent resource.

Contextual modeling is an alternative and complementary approach to current UML-based business modeling techniques.

Bruce Powel Douglass presents a strategy for managing your modeling artifacts in the article "Organizing Models the Right Way" (August 2000, Section 2.1.3). As you see in Figure 2.2, modeling can become quite sophisticated, particularly if the system you are modeling is large and/or complex. Therefore, to be successful, you need a strategy for organizing your modeling efforts. This is true of all modeling workflow efforts, including the Requirements, Business Modeling, and the Analysis and Design workflows. This article deftly addresses issues such as how application scale, team size, business culture and developer workflow affect model organization and the best way to work with models for your specific project? By answering these questions, Douglass describes a collection of strategies for organizing your model, strategies based on the type of project, your organization culture, and your software process — strategies that provide the infrastructure for team collaboration. The primary benefit of model organization is that it allows many team members to contribute to the model without corrupting it or losing previous changes. The approach that you choose to organize your models should reflect your unique environment. Before you simply accept these model organizations as they are, evaluate your own business culture, processes, and project characteristics to see which, if any, of these existing strategies best fits your project — or how a strategy might be modified to make your environment and product vision even better.

An important skill for a business modeler to have is the ability to recognize problems/issues that are similar to those occurring in a different context. When you recognize that you have seen a problem before, you are in a position to reuse a known/existing solution to that problem. The problem is how do you learn about existing solutions to common problems? The solution is to learn about patterns, a topic that Luke Hohmann, author of *Journey of the Software Professional* (1996), overviews in "Getting Started with Patterns" (February 1998, Section 2.1.4). Patterns[1] document the wisdom of other developers' proven solutions in such a way that you can build on this wisdom through your own skill and experience. It is this sharing of wisdom that makes patterns so incredibly powerful, however, patterns aren't a silver bullet that can solve every problem in software development. For additional examples of patterns, refer to "Requirements Engineering Patterns" in Section 3.5.3.

Patterns enable you to reuse the wisdom and experiences of other software professionals.

Class Responsibility Collaborator (CRC) modeling has been a primary modeling technique for understanding the business context for a system since it was first introduced by Kent Beck and Ward Cunningham (1989). Although not part of the UML, as you saw in Figure 2.2, CRC models are an important part of both object-oriented development methodologies (Ambler, 2001) where they are used for context/conceptual modeling, as well as the eXtreme Programming (XP) software process (Beck, 2000). In "CRC Cards for Analysis" (October 1995, Section 2.1.5) Nancy Wilkinson overviews the CRC modeling technique in

---

1. Visit *The Process Patterns Resource Page* http://www.ambysoft.com/processPatternsPage.html for more information regarding patterns.

an excerpt from her book *Using CRC Cards* (1995). Although the focus of the article is using CRC cards as an analysis technique, the reality is that it's a valid technique for business modeling, helping you to understand the problem domain better and providing a technique for working closely with your users. The article describes the technique and even discusses how to work through scenarios to evolve the CRC cards, a technique similar to the use case scenario testing presented in Section 4.4.4. Wilkinson also discusses how to use CRC models to promote reuse of domain classes within your organization, a concept appropriate to the Infrastructure Management workflow (see *The Unified Process Elaboration Phase* and *The Unified Process Construction Phase*, the second and third books in this series, for details regarding this workflow) discussing how to take advantage of existing domain classes as well as identify classes that you may be able to generalize to make reusable.

2.1   The Articles

2.1.1 "How the UML Models Fit Together" by Scott Ambler
2.1.2 "Data-Based Design" by Hugh Beyer and Karen Holtzblatt
2.1.3 "Organizing Models the Right Way" by Bruce Powel Douglass
2.1.4 "Getting Started with Patterns" by Luke Hohmann
2.1.5 "CRC Cards for Analysis" by Nancy Wilkinson

2.1.1 "How the UML Models Fit Together"

by Scott Ambler

At this point in its evolution, the Unified Modeling Language (UML) is recognized as a modeling language, not as a methodology or method. The difference is that a modeling language is more of a vocabulary or notation (in the case of UML, a mostly graphical notation) for expressing underlying object-oriented analysis and design ideas. A method or process consists of recommendations or advice on how to perform object-oriented analysis and design.

This article explains the UML notation, describes each UML modeling diagram, provides an example where appropriate, and indicates when the diagram can be used. I'm not going to go into enough detail to teach you how to actually model with UML; instead, at the end of the article, I'll outline a simple and useful process that uses UML.

The UML diagrams I'll cover will include the following:

* Use case diagrams
* Class diagrams
* Sequence diagrams
* Component diagrams
* Deployment diagrams
* Statechart diagrams
* Collaboration diagrams

Defining Use Cases and Use-Case Diagrams

A use case scenario is a description, typically written in structured English or point form, of a potential business situation that an application may or may not be able to handle. A use case describes a way in which a real-world actor — a person, organization, or external system — interacts with your organization. For example, the following would be considered use cases for a university information system:

- Enroll students in courses
- Input student course marks
- Produce student transcripts.

To put your use cases into context, you can draw a use case diagram, as shown in Figure 2.3. Use case diagrams are straightforward, showing the actors, the use cases with which they're involved, and the boundary of the application. An actor is any person, organization, or system that interacts with the application but is external to it. Actors are shown as stick figures, use cases are ovals, and the system is a box. The arrows indicate which actor is involved in which use cases, and the direction of the arrow indicates flow of information (in the UML, indicating the flow is optional, although I highly suggest it). In this example, students are enrolling in courses via the help of registrars. Professors input and read marks, and registrars authorize the sending out of transcripts (report cards) to students. Note how for some use cases, more than one actor is involved, and that sometimes the flow of information is in only one direction.

Figure 2.3   An example of a use case diagram.

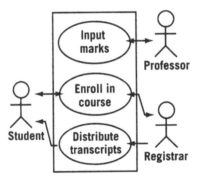

The combination of use cases and their corresponding use case diagram is referred to as a use case model. Use case models are often accompanied by a glossary describing the terminology used within them. The glossary and use case model are important inputs for the development of class diagrams.

Class Diagrams

Class diagrams, formerly called object models, show the classes of the system and their interrelationships (including inheritance, aggregation, and associations). Figure 2.4 shows an example of a class diagram, using the UML notation that models my Contact-Point analysis pattern (Ambler, 1998a). Class diagrams are the mainstay of object-oriented modeling and are used to show both what the system can do (analysis) and how the diagram will be built (design).

Figure 2.4   A class diagram representing the Contact-Point analysis pattern.

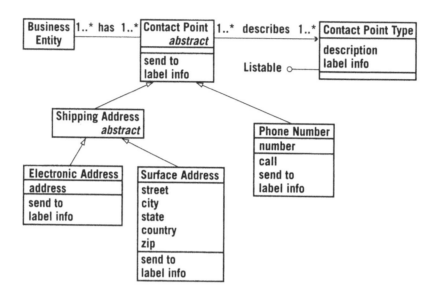

Class diagrams are typically drawn by a team of people led by an experienced object-oriented modeler. Depending on what is being modeled, the team will be composed of experts on the subject who supply the business knowledge captured by the model and any other developers who provide input into how the application should be designed. The information contained in a class diagram directly maps to the source code that will be written to implement the application. Therefore, a class diagram must always be drawn for an object-oriented application.

Classes are documented with a description of what they do, methods are documented with a description of their logic, and attributes are documented by a description of what they contain, their type, and an indication of a range of values (if applicable). Statechart diagrams, covered later, are used to describe complex classes. Relationships between classes are documented with a description of their purpose and an indication of their cardinality (how many objects are involved in the relationship) and their optionality (whether or not an object must be involved in the relationship).

Sequence Diagrams

A sequence diagram, formerly called an object-interaction or event-trace diagram, is often used to rigorously define the logic for a use case scenario. Because sequence diagrams look at a use case from a different point of view from which it was originally developed, it is common to use sequence diagrams to validate use cases. Figure 2.5 shows an example of a sequence diagram. Depending on your modeling style, it's also common to use sequence diagrams during design to understand the logic of your application. This is usually done by a group of developers, often the programmers responsible for implementing the scenario, led by the designer or architect for the project.

Figure 2.5   **A sequence diagram for transferring funds from one account to another.**

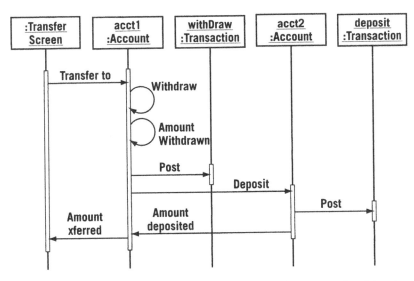

Traditional sequence diagrams show the types of objects involved in the use case, the messages they send each other, and any return values associated with the messages (many modelers choose to show return values only when it isn't obvious what is being returned by the method). Objects (instances) in the UML are shown underlined to distinguish them from classes. For large applications, it's common to show the components and use cases in addition to objects across the top of the diagram. This makes sense because components are really just a reusable object and use cases are implemented as objects whose role is to encapsulate the process and control the flow defined by the use case. The basic idea is that a sequence diagram visually shows the flow of logic of a use case, letting you document and reality check your application design at the same time. The boxes on the vertical lines are called method invocation boxes, and they represent the running of a method in that object.

Sequence diagrams are a great way to review your work, as they force you to walk through the logic to fulfill a use case scenario. They also document your design, at least from the use case point of view. By looking at what messages are being sent to an object, component, or use case, and by looking at roughly how long it takes to run the invoked method, sequence diagrams give an understanding of potential bottlenecks, letting you rework your design to avoid them. When documenting a sequence diagram, it's important to maintain traceability to the appropriate methods in your class diagram(s). The methods should already have their internal logic described as well as their return values (if they don't, you should document them immediately).

Component Diagrams

Component diagrams show the software components that make up a reusable piece of software, their interfaces, and their interrelationships. For the sake of this article, I take the component/package approach to architectural reuse. This approach is described in

Ivar Jacobson's book *Software Reuse* (Addison Wesley, 1997). From this vantage, a component may be any large-grain item — such as a common subsystem, an executable binary file, a commercial off-the-shelf system (COTS), an object-oriented application, or a wrapped legacy application — that is used in the day-to-day operations of your business. In many ways, a component diagram is simply a class diagram at a larger, albeit less-detailed, scale.

Figure 2.6 shows an example of a component diagram that models the architectural business view of a telecommunications company. The boxes represent components, in this case either applications or internal subsystems, and the dotted lines represent dependencies between components. One of the main goals of architectural modeling is to partition a system into cohesive components that have stable interfaces, creating a core that need not change in response to subsystem-level changes. Component diagrams are ideal for this purpose.

Figure 2.6 **A component diagram for the architectural business view of a telecommunications company.**

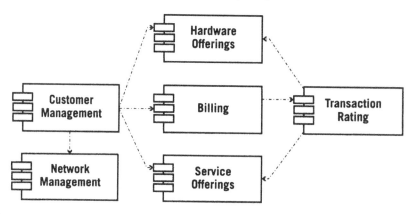

Each component within the diagram will be documented by a more detailed component diagram, a use case diagram, or a class diagram. In the example presented in Figure 2.6, you'd probably want to develop a set of detailed models for the component Customer Management because it's a reasonably well-defined subset. At the same time, you would draw a more detailed component diagram for Network Management because it's a large and complex domain that must be broken down further.

This approach to component diagrams is contrary to what the UML document currently suggests. The current use for component diagrams proposed by the UML document is that it is an implementation diagram that shows the granular components, perhaps ActiveX components or Java Beans, used during implementation. Although these kinds of components are important, this is a low-level approach to reuse. Bigger payback is achieved from taking an architectural approach to development. The manner in which I suggest using component diagrams supports this approach. Besides, you still have deployment diagrams to document your implementation decisions.

Deployment Diagrams

Deployment diagrams show the configuration of run-time processing units, including the hardware and software that runs on them. Figure 2.7 shows an example of a deployment diagram that models the configuration of a three-tiered client/server customer service application. Notice the similarity of the notation for both deployment and component diagrams. Deployment diagrams are reasonably simple models that are used to show how the hardware and software units will be configured and deployed for an application.

Figure 2.7 **A deployment diagram for a three-tiered client/server application.**

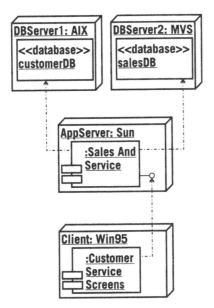

For each component of a deployment diagram, you'll want to document the applicable technical issues such as the required transaction volume, the expected network traffic, and the required response time. Further, each component will be documented by a set of appropriate models. For example, the databases will be described with data models, the application server will be described with a component diagram or class diagram, and the customer service screens will at least be documented by an interface-flow diagram and a prototype.

Statechart Diagrams

Objects have both behavior and state; in other words, they do things and they know things. Some objects do and know more things, or at least more complicated things, than other objects. Some objects are incredibly complex, so statechart diagrams are drawn to describe how they work, as shown in Figure 2.8.

Figure 2.8 **A statechart diagram for a bank account.**

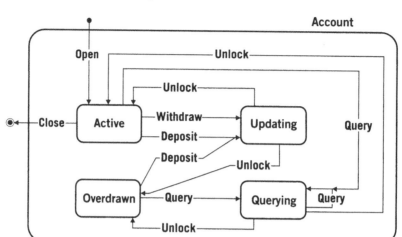

Figure 2.8 shows the statechart diagram for a bank account. The rectangles represent states that are stages in the behavior of an object. States are represented by the attribute values of an object. The arrows represent transitions, which are progressions from one state to another and are represented by the invocation of a method on an object class. Transitions are often a reflection of business rules. There are also two kinds of pseudo states: an initial state, in which an object is first created, and a final state that an object doesn't leave once it enters. Initial states are shown as solid circles, and final states are shown as an outlined circle surrounding a solid circle. In Figure 2.8, you can see that when an account is active, you can withdraw from it, deposit to it, query it, and close it.

Collaboration Diagrams

Unlike some notations that show both state and behavior on class diagrams, the UML models behavior using collaboration and sequence diagrams (this is called dynamic modeling). The basic difference between the two approaches is that UML class diagrams are used for static modeling, they don't include messages because messages tend to clutter your class diagrams and make them difficult to read. Because UML class diagrams don't show the message flow between classes, a separate diagram, the collaboration diagram, was created. Collaboration diagrams show the message flow between objects in an object-oriented application and imply the basic associations between objects.

Figure 2.9 presents a simplified collaboration diagram for a university application. The rectangles represent the various objects and the roles they take within your application, and the lines between the classes represent the relationships or associations between them. Messages are shown as a label followed by an arrow indicating the flow of the message, and return values are shown as labels with arrow-circles beside them. In Figure 2.9, there are instances of the Seminar and Enrollment classes, open and display info messages, and seats, which is a return value (presumably, the result of sending the message max seats to Course).

Figure 2.9   **A collaboration diagram for a simple university application.**

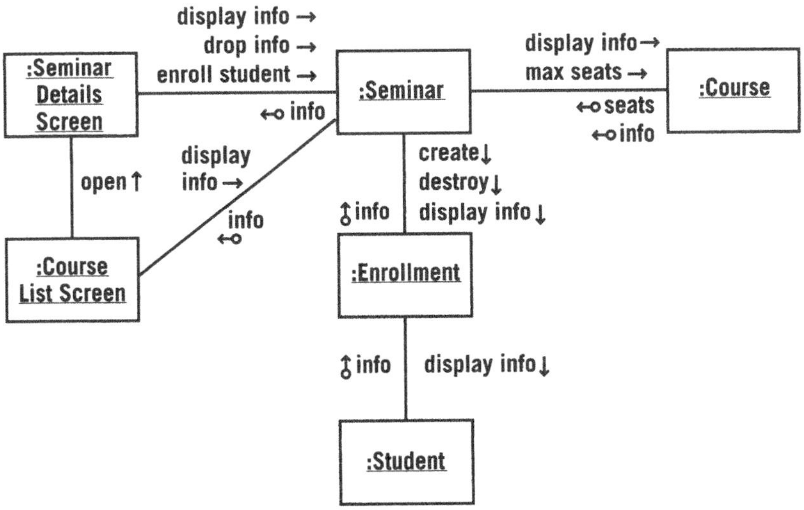

Collaboration diagrams are usually drawn in parallel with class diagrams, especially when sequence diagrams haven't been developed for your application. You can use collaboration diagrams to get the big picture of the system, incorporating the message flow of many use case scenarios. Although you can indicate the order of message flow on a collaboration diagram by numbering the messages, this typically isn't done because sequence diagrams are much better at showing message ordering.

How the UML Diagrams Fit Together

In Figure 2.10, the boxes represent the main diagrams of the UML, and the arrows show the relationships between them. The arrowheads indicate an "input into" relationship. For example, a use case diagram is an input for a class diagram. Figure 2.10 provides insight into one of the fundamentals of object-oriented modeling. The relationships between the various models of the UML are reflections of the iterative nature of object-oriented modeling.

Figure 2.10   **UML modeling techniques from an iterative point of view.**

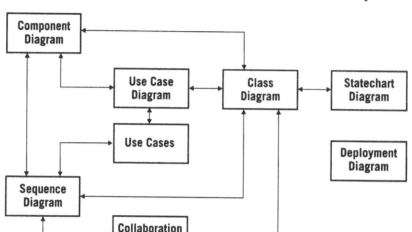

Figure 2.11 shows a slightly different view of the construction process, a serial one, in which the lines between the boxes represent a "documented by" relationship instead of the "input into" relationship shown in Figure 2.10. For example, Figure 2.11 shows that a statechart is used to document a class diagram (actually, a statechart would be used to document a single class shown on a class diagram). Similarly, components on a component diagram may be documented by another component diagram, a class diagram, or a use case model. Figure 2.11 shows that source code is used to document the classes on a class diagram.

Figure 2.11   **UML modeling techniques from a serial point of view.**

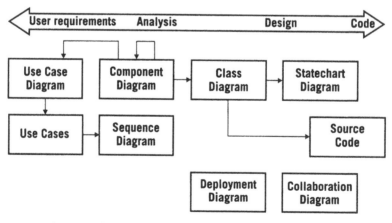

An interesting feature of Figure 2.10 and Figure 2.11 is they illustrate that the object-oriented modeling process is serial in the large and iterative in the small. The serial nature is exemplified by Figure 2.11 and the iterative nature by Figure 2.10.

At the beginning of large object-oriented projects, you typically start with requirements-gathering techniques such as use cases, then move forward into analysis and design techniques such as class diagrams, and finally progress to spending the majority of your efforts coding. Although you are iterating through all of these techniques, and you may very well write some code at the beginning of your project, the fact remains that you generally need to identify requirements first, then analyze them, then design your software, then code it. In other words, object-oriented development is serial in the large and iterative in the small. Let's spend some time discussing a simplified modeling process using the UML diagrams that we've already discussed.

Class-Driven Application Modeling

Figure 2.12 shows a modeling process that I call "class-driven" because requirements are captured via class responsibility collaborator (CRC) modeling. CRC modeling is a technique in which a group of domain experts, led by a facilitator, defines the requirements for an application on standard index cards organized in three sections: the name of the class, its responsibilities (what the class knows and does), and the other classes with which the class collaborates to fulfill its responsibilities.

Figure 2.12   A class-driven modeling process.

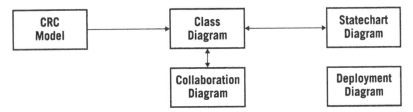

I take this approach in environments where I have access to users who do the work day in and day out. CRC modeling is a simple technique that you can teach in a few minutes. It's non-technical and unthreatening to people who are afraid of being computerized out of a job, and it fits well with class diagramming as it models many of the same concepts. Although the CRC modeling process isn't defined by the UML, it's a well-recognized modeling technique.

An interesting difference between a class-driven modeling process and one that revolves around use cases (see Figure 2.3) involves the use of collaboration diagrams and not sequence diagrams. Use case and class diagrams show similar information — the dynamic behavior of your software — although in a slightly different manner. Sequence diagrams are excellent for documenting the logic of a single use case, whereas collaboration diagrams are more appropriate for showing the logic of what may be several use cases. When you take an approach in which use cases are used to define requirements, you may be tempted to use sequence diagrams. With a class-driven approach, you're tempted to use collaboration diagrams because they're closer in a notational sense to class diagrams than sequence diagrams are.

I prefer to use collaboration diagrams when I need to show asynchronous messaging between objects, and I prefer sequence diagrams for synchronous logic. Yes, the UML provides mechanisms for showing asynchronous messaging on sequence diagrams, and synchronous messaging on collaboration diagrams, but frankly the mechanisms are clunky at best. Each UML diagram has its strengths and weaknesses, and this is a perfect example of when you want to use one diagram over another.

The application and system modeling process shown in Figure 2.12 is one of many that your organization can employ. My general rule of thumb is to develop a diagram if it helps me understand what I'm building, but if it isn't going to help, I don't develop it.

2.1.2 "Data-Based Design"

by Hugh Beyer and Karen Holtzblatt

The first steps toward creating a usable system are making sure it addresses the customers' needs and that it works the way they expect. In this article, we'll show you how to achieve these goals.

When a development team creates a system, it is actually inventing a new way for people to work. A software product must do this to make a splash in the market. A customer system, therefore, seeks to simplify and streamline a work task with the appropriate automation. Even the smallest tool with the most limited effect must integrate seamlessly with the larger work practice. In every case, what makes a system interesting to its users is the new work process it enables.

Use cases, storyboards, and scenarios are all ways of representing the new work practice and figuring out how technology will support it. But no matter how the practice is captured, a design team needs to decide what it will be — what new structure of tasks and activities will improve work practices.

Contextual Design is a customer-centered approach to software design that grounds design decisions in customer data. It involves using team processes to gather field data about customers, representing the work of individuals and whole customer populations in work models, using these models to reinvent the way people work, and designing and iterating the system with customers. The crux of this process is understanding and reinventing the way people work based on field data. This gives you license to innovate because you'll know people's hidden motives, their strategies, and the things they care about. Your invention will take account of people's real needs.

In this article we'll discuss how to see common aspects of work across a customer population, no matter how diverse and idiosyncratic it may seem at first, and how to use this detailed customer knowledge to drive innovative designs.

Contextual Design uses five kinds of work models to reveal common aspects of work: flow models showing communication and coordination between people; context models showing the cultural and organizational environment in which work happens; physical models showing the aspects of the physical environment affecting the work; sequence models showing the ordered work steps across time; and artifact models showing the structure and intent of artifacts in the work.

We'll focus on sequence models for this article. From this foundation you can develop use cases — though we'll only sketch out the procedure, it's straightforward once you have a solid representation of the work and a vision for how to change it. You can also continue the design process with storyboards, prototyping, or other activities within Contextual Design.

Creating One View of a Customer Population

It's remarkable that systems can be built to support large numbers of people at all. People don't coordinate the way they work across industries, and so they seldom work in consistent ways. Even in a single department, people develop idiosyncratic ways of doing their jobs. But systems are made for whole customer populations, the intended users of a system in the market to which a product is sold, or in an organizations' departments.

Any system imposes work practices on its users — it structures work and interacts with it in many complex ways. Since a system always structures users' work, and since people don't coordinate to work consistently, why should a single system work for them all?

We take it for granted that products and systems can be built and will be successful with all their disparate users. If a system can address the needs of a whole customer population, it's because aspects of work hold across all customers: common structure, strategy, and intent. A design responds to this common work structure while allowing for individual variation among customers.

But how can we discover common aspects of work among all the surface differences? How do we represent the common aspects of work so a whole design team can respond to them?

The first step is to discover how individual customers work on a day-to-day basis. Contextual Inquiry, our field data gathering technique, does this by sending designers to the field to observe customers doing their day-to-day work. By discussing it with them, designers discover the underlying intents, strategies, and concepts driving the work.

From interviews such as these, designers create work models to represent individual customers. Each type of work model shows a different aspect of work relevant to design. To see the work of the customer population, these models of individuals are consolidated to produce a single model of each type that reveals common patterns of work.

Consolidation is an inductive process of building a larger structure up from real cases, not from theoretical or abstract arguments. It's too easy to make decisions about work based on what you think must be true, not on what's real for your customers.

Consolidating sequence models follows this pattern. First, we create individual sequence models that describe one real instance of work, showing how one person accomplishes a task. Sequence models show the trigger that initiated the task, the steps or actions that accomplish it, and the intents or motives for doing it.

From the individual sequence models, we derive a consolidated sequence model. This model shows designers the detailed structure of a task. It reveals the common strategies and all the different intents across a customer population that must be supported by the system or rendered unnecessary. The consolidated sequence model is the designer's map showing how a user approaches a task. It's the starting point for any use case or storyboard.

Identifying User Activities

The first step in consolidating sequences is to identify the users' "activities." The sequences in Figure 2.13 show how we observed two system managers as they diagnosed problems. Skimming Manager 1's sequence, we see that an automated process notifies him that something is wrong, he pokes around looking for problems, then he calls for help. These immediately become potential activities: notify, diagnose, get help. Manager 2 is notified by a person, she pokes around on the hardware until she recognizes the problem is something AT&T has to fix, and then she calls them for help. Again we see the basic structure of activities: notify, diagnose, get help. (We'll save the rest of the sequences for later.) For now, we'll match the steps in the sequence that initiate a new activity, as shown in Figure 2.14.

Figure 2.13   Individual sequences for diagnosing system problems.

MANAGER 1: INDIVIDUAL SEQUENCE	MANAGER 2: INDIVIDUAL SEQUENCE
Fix All-in-1	Fix router problem
Trigger: Watcher sends mail that the All-in-1 (A1) mail system isn't working	Trigger: Person walks into office to report problem—can't access files on another machine
Logs on to failing system to search for problem	Goes into lab to look at equipment
Discovers the A1 mail process has crashed (ACCVIO)	Flicks switches to do loop-back tests, isolating wire, MUX, router
Looks at the time of the crash: only this morning	Determines problem—bad router
Tries to restart the process by hand	Calls AT&T people in second organization
Process won't restart	Does something else while waiting for AT&T to show up
Looks at the process log: can't tell why it won't start	AT&T comes to look at problem
Calls expert backup group for help	Looks in book to tell which wire is which; it shows which nodes are on which wires, and which wire goes to which router
Asks them to log into the system and look at the problem	Tells AT&T people which router is at fault and which wire it's on
Keeps looking for problem in parallel	AT&T people fix problem
Searches for problem	Logs problem and fix
Discovers that process can't create a needed directory	Done
Tries to create directory manually	
[Looks to see if directory was created]	
Can't create directory—disk is too fragmented	
Calls expert backup to explain —types and talks on speaker phone at the same time	
Discusses problem—agrees on the exact procedure to follow	
Implements fix	
Writes mail to users describing changes that affect them	
Done	

Figure 2.14   Individual work activities.

MANAGER 1: INDIVIDUAL SEQUENCE	MANAGER 2: INDIVIDUAL SEQUENCE
Notify	**Notify**
Trigger: Watcher sends mail that the All-in-1 (A1) mail system isn't working	Trigger: Person walks into office to report problem—can't access files on another machine
Diagnose	**Diagnose**
Logs on to failing system to search for problem	Goes into lab to look at equipment
Discovers the A1 mail process has crashed (ACCVIO)	Flicks switches to do loop-back tests, isolating wire, MUX, router
Looks at the time of the crash: only this morning	Determines problem—bad router
Tries to restart the process by hand	
Process won't restart	
Looks at the process log: can't tell why it won't start	
Get help	**Get help**
Calls expert backup group for help	Calls AT&T people in second organization

The first step of a sequence is the *trigger* that initiates it. Manager 1's actions were triggered by an e-mail; Manager 2's actions were triggered by a person. The first step of the consolidated sequence represents the different triggers that can initiate it. In this case, we list the two triggers as alternative ways to find out about the problem.

In other cases, a different trigger might change the strategy. System managers who are notified of a problem by a help desk may go right into hypothesis testing, whereas if they are notified of the problem by a report from an automated process, they may need to find out more about it first. When this happens, we keep the triggers separate in the consolidated sequence to show how they initiate different work strategies.

The next set of steps in both individual sequences contributes to diagnosing the problem. Our task is to match steps that accomplish the same thing in each instance and define "abstract steps" for them, as shown in Figure 2.15. The abstract steps summarize the action of the individual sequence steps. We don't yet know exactly how the steps match up; we only know that they all have to be sorted out before getting to the steps in which both managers call for help. The first step in each case positions the user in the right place to diagnose the problem: either logging in or going to the computer lab. Logging in or going to the computer lab is the detail unique to the instance. This is our next abstract step, as shown in Figure 2.16.

Figure 2.15   Creating abstract steps.

ABSTRACT STEP	MANAGER 1	MANAGER 2
Trigger: Find out about problem —Automated procedure —Someone reports problem	Trigger: Watcher sends mail that the All-in-1 (A1) mail system isn't working	Trigger: Person walks into office to report problem—can't access files on another machine

Figure 2.16   Abstract step for diagnosing a problem.

ABSTRACT STEP	MANAGER 1	MANAGER 2
Go to the place where the problem can be solved (physically or logically)	Log on to failing system to search for problem	Goes into lab to look at equipment

Next, both managers try different things on the system until the problem is identified. You'll notice in Figure 2.14 that "Discover the A1 mail process has crashed" and "Determine problem — bad router" both seem to mark the point at which the user identifies the problem. Manager 2's sequence has a step where she flicks switches and runs tests to determine the problem. The team who wrote Manager 1's sequence didn't record such a step, but it's implied by "Discover the A1 mail process has crashed." But as Manager 1's sequence indicates, all that's been discovered so far is why the symptom is happening — the underlying problem (a full disk in Manager 1's case) may not have been determined yet. The consolidation is shown in Figure 2.17.

Figure 2.17   Consolidation of problem identification.

ABSTRACT STEP	MANAGER 1	MANAGER 2
Execute commands and tests on suspect system to identify anomalous behavior	Do something to discover the A1 process isn't running	Flicks switches to do loop-back tests, isolating wire, MUX, router
Determine cause of symptoms	Discover the A1 mail process has crashed (ACCVIO)	Determine problem— bad router

At this point, the two managers diverge in their strategies. Manager 1 tries to fix the problem. Manager 2 decides she must call AT&T to do the fix. Neither manager's decisions are written explicitly, but both are implied by their actions. So the abstract steps branch to account for the two cases. The consolidation will incorporate both cases, so it collects all possible ways of doing the work that a design team needs to respond to. This consolidation is shown in Figure 2.18.

Figure 2.18 Consolidation of divergent strategies.

ACTIVITY	ABSTRACT STEP	MANAGER 1	MANAGER 2
Diagnose problem	Estimate impact of problem	Look at the time of the crash: only this morning	
	Decide if I can fix the problem	(Decide to fix)	(Decide AT&T has to fix)
	If I decide I can fix it:		
	Attempt fix	Try to restart the process by hand	
	See if fix worked	Process won't restart	
	If fix didn't work, try to figure out why	Look at the process log: can't tell why it won't start	
Get help	Decide I can't fix it, call for help	Call expert backup group for help	Call AT&Tpeople in second organization

This process repeats until the whole sequence is consolidated. We identify the sections of the sequences that are similar, match up individual steps, and name abstract steps for them. Either after a whole activity or at the end of the sequence, we step back and ask the intent of each step. What are the obvious and the nonobvious reasons for doing the step? Any step may have more than one intent, and high-level intents may cover a whole set of steps. It's easy to identify and support the main intent of the sequence. It's harder to see all the additional, secondary intents that are also important to the customer. We decide what they are and write them down. If we aren't sure of an interpretation, we talk to the interviewer or check back with the user. The result, for the sequences we've been consolidating, is shown in Figure 2.19.

Of course, real teams consolidate three or four actual sequences at once, not just two. The first cut at abstract steps is correspondingly more robust. Once the initial set of sequences has been consolidated, the rest of the sequences are compared with the consolidated sequence and used to extend it. Incorporating more sequences will add additional steps, show new strategies, and provide more alternatives for steps that are already represented.

Consolidated sequence models define a backbone into which new variations can be incorporated and accounted for. In our example, it's not hard to see how a new trigger or new step in diagnosing a problem could be incorporated into the structure we developed. Armed with this knowledge, designers can structure their systems to reflect the structure of the task. This is where redesigning the work comes into play.

Figure 2.19   Final consolidation for two users' work processes.

ACTIVITY	INTENT	ABSTRACT STEP
Find out about problem	Learn about problems quickly	Trigger: Find out about problem
	Discover problems before users do	—Automated procedure —Someone reports problem
	Provide quick response	
Go to problem location	Make it possible to do diagnosis and take action	Go to the place where the problem can be solved
Diagnose problem	Find cause of problem	Execute commands and tests on suspect system to identify anomalous behavior
		Determine cause of symptoms
	Decide who's been affected	Estimate impact of problem
	Decide if any additional action should be taken to notify people of status	
Fix problem	Fix the problem at once	If I decide I can fix it: 　　Attempt fix
		See if fix worked
		If fix didn't work, try to figure out why
Call on help	Get the people involved who have the authority or the knowledge to fix the problem	Decide I can't fix it, call on help
	Ensure problem gets fixed, even if not my job	

Redesigning Work

The consolidated sequence model reveals the structure of a task. This structure becomes the basis from which you design. In the design process, teams create a consolidated model for each of the five perspectives on work, including consolidated sequence models for each major task their system will support. These models guide how the task will be reinvented in the new system. They show what people care about, how they approach the task, the strategies they use, and the problems to overcome. Your job is to reinvent the work practice by taking advantage of technology and considering how to respond to the strategy as it has been revealed. Following are some points to remember:

Address the intents.   Every consolidated sequence has a primary intent — the reason why the task was worth doing in the first place. Every consolidated sequence also has numerous secondary intents that are accomplished along the way. (Make sure I don't do things I'm not supposed to, for example.) The first question to ask of a sequence model is whether the primary intent really needs to be met. If you can redesign the work so the whole task becomes unnecessary, you've simplified the job.

But before you can eliminate the whole task, you must look at all the individual intents that the user accomplishes along the way. The most typical failure of a process reengineering effort is to make it impossible to accomplish an important intent that was never recognized in the redesign. When you walk through the intents of a sequence, some will support the primary intent of the sequence — if you eliminate the need for the task, these intents become unnecessary. But you will discover that others have nothing to do with the primary intent of the sequence. If you eliminate the whole sequence, you must find another way for the customers to accomplish these secondary intents or your system will fail.

If you choose to keep the sequence, every intent is an opportunity for redesign. Each intent indicates something that matters to the work — if you can provide a way to achieve them more directly, you can simplify the work. When redesigning how users accomplish an intent, treat it just as you treat the overall intent of the sequence — look at the part of the sequence you are designing away, and make sure your customers can still accomplish all the intents that matter to them.

Support the strategies. The consolidated sequence models show the different strategies that customers use to accomplish work tasks. Your system needs to recognize the strategies and support them or introduce a new way of working that supplants one or more of them. If you choose the latter option, then you have to account for the underlying characteristics driving customers to choose the strategy you are eliminating. You might decide that system managers who continue to work on a problem after turning it over to their backup experts are wasting their time. But if part of the reason they do this is to save face — to prove that even though they had to ask for help they are still experts and have just as good a chance of finding the problem as the people they called in — you won't be able to eliminate this behavior. You'd do better to recognize and allow for it.

Simplify the steps. After you've decided that your system must support the sequence, it becomes your guide to the structure of the task. Look at the structure of the sequence to reveal the issues for your design. For example, one of the activities in the above consolidated sequence is "Go to problem location." In your new system, will managers continue to walk to the place physically? If so, how will you help them take information about the problem with them? If not, how can you get them the diagnosis information they need in their office? Can you eliminate the activity by providing all the information they need with the initial notification?

The triggers show how to alert the user that something needs to be done. Pay attention to the style of the trigger — is it noisy or quiet? Is it appropriate to the work being triggered? Does it work, or is it a nuisance? In our example, is it a problem for people to drop in to notify the manager that something is wrong? Does it disrupt work, or is it desirable to have some human interaction? Don't try to create entirely automatic systems if your customers like the human aspects of their work.

Finally, look at the steps themselves. Can you simplify them? Where the customer currently takes several steps, can you automate them down to one? Where a step is currently difficult, can you make it easy? Maybe the best thing you can do for system managers is to make running diagnostic tests easier.

Looking at your customers' work in this way gives you a strong basis for deciding how to reinvent the way they work. Sequence models capture the new work practice and show how the new system integrates with the rest of the work. This gives you the information you need to decide what work structure to build into a use case. Or you can follow Contextual Design and work out the high-level structure of the system before capturing specific behavior in use cases. Whichever way you go, you have concrete customer data to guide you.

Design without customer data is hard. Without detailed data about how people work, the only rationale for one design decision over another is who can make the better (or louder) argument. However, a design can be built on and guided by concrete customer data that is collected from individuals and consolidated to show the work of the customer population. This approach to design makes it easier to agree and more likely that the design will meet customer needs. You can't ask much more of the front-end design process.

2.1.3 "Organizing Models the Right Way"

by Bruce Powel Douglass

In many of the places where I work and consult, the organization of the developing system model is a significant concern. How do application scale, team size, business culture and developer workflow affect model organization, and what is the best way to work with models for your specific project? This is important for all software projects because it provides the infrastructure for team collaboration. Good model organization is even more crucial for achieving reusability in developing frameworks and components.

Before we begin, I'm going to make some assumptions about your working environment. First, you have a model. That is, you're going to use the UML or some other language to construct a visual rather than a code-based model of your system (although many of the organizing principles I'll discuss apply equally well to code-based systems). Second, you're using a tool to manage the repository of your model's semantic elements. Third, not only are there several members on your team, but there are also dozens or even hundreds of colleagues who need to simultaneously use and contribute to the model. Last, I assume that you understand the basics of configuration management (CM) and have the infrastructure for it in place.

Why Organize Your Model?

In simple systems, you can basically do anything you like with the model and still succeed. Once you have a system complex enough to require teams of people working in tandem, however, it begins to matter what you do and how you do it. The primary benefit of model organization is that it allows many team members to contribute to the model without corrupting it or losing previous changes. Further, it lets team members use parts of the model that lie outside of their responsibility, provides for an efficient build process, helps developers locate and work on various model elements, and, finally, facilitates reuse of selected pieces of the system, like components.

You might think that configuration management solves the logistical problems of contributing to and using aspects of a common model. This is only partially true. CM does provide locks, ensuring that only one worker can own the "write token" to a particular model element and that other users can only have a "read token." However, this is similar to saying that C solves all your programming problems because it provides assignment, branching and looping. CM does not say anything about what model elements ought to be configuration items (CIs); certain usage policies apply only if that model element is a CI. Effective model organization uses the CM infrastructure, but provides a set of higher-level principles for effectively using the model.

For example, it would be extremely awkward if the entire model was the single CI. Then only one person could update the model at a time. The other extreme would be to make every element — every class and use case — a separate CI. Again, in simple systems, this can work because there are only a few dozen model elements, so it is not too onerous to explicitly check out each element. However, this does not scale well, even to medium-sized systems. I would hate to have to individually list 30 or 40 classes when I wanted to work on a large collaboration realizing a use case.

The UML provides an obvious organizational unit for a CI — the "package." A UML package is essentially a bag into which we can throw semantic model elements such as use cases, classes, objects and diagrams. However, the UML does not provide any criteria for what should go into what package versus another. So, while we might want to make packages CIs in our CM system, this begs the question as to what policies and criteria we should use to decide how to organize our packages — what model elements should go into one package versus another.

A simple solution would be to assign one package per worker. Everything that Sam works on is in SamPackage; everything that Julie works on is in JuliePackage. For very small project teams, this is, in fact, a viable policy. But again, this begs the question of what Sam should work on versus Julie. It can also be problematic if Susan wants to update a few of Sam's classes while Sam is working on some others in SamPackage. Further, this scheme adds the artificial dependencies of the model structure on the project team organization. This will make it more difficult to make changes to the project team (say, to add or remove workers) and will also limit reusability.

Watch the Workflow

In addition to configuration items, it also makes sense to examine the development workflow before organizing the model. Common workflows in the software life cycle include:

- gathering requirements (detailing use cases, scenarios, statecharts or activity diagrams),
- realizing requirements with analysis and design elements (elaborating and refining collaborations, detailing an individual class),
- designing the logical and physical architecture (working on a set of related classes from a single domain or working on a set of objects in a single run-time subsystem or component),
- constructing prototypes and testing (translating requirements into tests, building iterative prototypes from model elements), and

- planning (project scheduling, including work products from the model).

Use cases are central to gathering requirements, and are an obvious focal point for organizing the requirements and analysis model. For example, when working on related requirements and use cases, the developer typically needs access to one or more use cases and actors. When detailing a use case, a developer will work on a single use case and detailed views — a set of scenarios and often either an activity diagram or a statechart (or some other formal specification language). When elaborating a collaboration, the user will need to create a set of classes related to a single use case, as well as refining the scenarios bound to that use case. Packages can divide up the use cases into coherent sets (such as those related by generalization, <<includes>> or <<extends>> relations, or by associating with a common set of actors). In this case, a package would contain a use case and its actors, activity diagrams, statecharts and sequence diagrams.

The next workflow (realizing requirements) focuses on classes, which may be used in either analysis or design. A domain, as defined in the ROPES process (see my book *Doing Hard Time: Developing Real-Time Systems with UML, Objects, Frameworks and Patterns*, Addison-Wesley, 1999), is a subject area with a common vocabulary, such as device I/O, user interface or alarm management. Each domain contains many classes, and system-level use case collaborations will contain classes from several different domains. Many domains require rather specialized expertise, such as low-level device drivers, aircraft navigation and guidance, or communication protocols. From a workflow and logical standpoint, it makes sense to group such elements together because a single person or team will develop and manipulate them. Grouping classes by domains and having the domains be CIs may make sense for many projects.

If a class myClass *has some testing support classes, such as* myClass_tester *and* myClass_stub, *they should be kept together.*

Architecture

Architectural workflows also require effective access to the model. One approach is to break up the architecture into the logical (organization of types, classes and other design-time model elements) and physical aspects (organization of instances, objects, subsystems and other run-time elements). The logical architecture is often organized by domains, while the physical architecture revolves around components or subsystems. If you structure the model this way, then a domain, subsystem or component is made a CI and assigned to a single worker or team. If the element is large enough, then it may be further subdivided into subpackages of finer granularity based on subtopic within a domain, subcomponents or other criterion, such as team organization.

Testing workflows are often neglected in model organization — usually to the detriment of the project. Though testing teams require read-only access to the model elements, they nevertheless need to manage test plans, procedures, results, scripts and fixtures, often at multiple levels of abstraction. Testing typically occurs on primary levels: unit testing (often done by the person responsible for the particular model element or a peer), integration (internal interface testing) and validation (black-box, system-level testing).

Because unit-level testing consists primarily of white box, design or code-level tests and often uses additional model elements constructed as test fixtures, it makes sense to co-locate them with the corresponding parts of the model. So, if a class `myClass` has some testing support classes, such as `myClass_tester` and `myClass_stub`, they should kept together, either within the same package or in another if a peer will do the testing — so long as it is a different CI than the model elements under test.

Integration and validation tests are not as tightly coupled as unit level tests, but the testing team still may construct model elements and other artifacts to assist them. Because the creators of the model elements don't typically do these tests, independent access is required, and they should be in different CIs.

Efficiently constructing and testing prototypes is a crucial part of the development life cycle. This involves both tests against the architecture (integration) and against the entire prototype's requirements (validation). There may be any number of model elements specifically constructed for a particular prototype that need not be used anywhere else. It makes sense to keep these near that build or prototype. Store test fixtures to be applied to many or all prototypes in a locale that allows independent access from any given prototype.

Store test fixtures to be applied to any or all prototypes in a locale that allows independent access from any given prototype.

Four Organization Patterns

Now that we've reviewed the factors influencing model usage, let's look at four traditional ways of organizing models: by use cases, by frameworks, by logical models and by builds.

The simplest way to organize models is around use cases, breaking the system into several high-level packages (one for the system level and one for each use case). The advantages of this organization are its simplicity and the ease with which requirements can be traced from the high level through the realizing elements. Also, it is a comfortable arrangement for a small system (say, three to 10 use cases and one to six developers). However, this approach doesn't scale up to medium or large systems, and the model elements are difficult to reuse. Other disadvantages include the fact that the organization doesn't provide a place to put elements common to multiple use case collaborations, hence developers using this model tend to reinvent similar objects. Finally, there is no place to put larger-scale architectural organizations in the model, further limiting its scalability to large systems.

A framework-based model organization addresses some of the limitations of the use case based approach. It is still targeted at small systems, but it adds a framework package for shared and common elements. The framework package has subpackages for usage points (classes that will be used to provide services for the targeted application environment) and extension points (classes that will be subclassed by classes in the use case packages). It should be noted that there are other ways to organize the framework area that also work well. For example, frameworks often consist of sets of coherent patterns; the subpackaging of the framework can be organized around those patterns. While this is particularly apt when constructing small applications against a common framework, the scheme does hamper reuse in some of the same ways as the use case-based approach.

As I mentioned earlier, if the system is simple enough, workflow and collaboration can be on an ad hoc basis. A small application might be 10 to 100 classes realizing three to 10 use cases — on the order of 10,000 lines of code. By contrast, one of the characteristics of successful large systems (more than, say, 300 classes) is that architectural concerns play an important role in organizing and managing the application. In the ROPES process, there are two primary subdivisions of architecture: logical and physical (see "Components: Logical and Physical Models," in The Unified Process Construction Phase, Ambler and Constantine, 2000b) . The logical model organizes types and classes (things that exist at design time), while the physical model organizes objects, components, tasks and subsystems (things that exist at run time). When reusability of design classes is your goal, maintaining this distinction in the model organization is extremely helpful.

One of the characteristics of successful large systems is that architectural concerns play an important role in organizing and managing the application.

Logical Models

Large-scale systems require packages for the system, logical model, subsystem models and builds. The system package contains elements common to the overall system: system-level use cases, subsystem organization (shown as an object diagram) and system-level actors. The logical model is organized into subpackages called domains. Each domain contains classes and types organized around a single subject matter, such as user interface, alarms, hardware interfaces, bus communication and so on. Domains have domain owners — those workers responsible for the content of a specific domain. Every class in the system ends up in a single domain. Class generalization hierarchies almost always remain within a single domain, although they may cross package boundaries within a domain.

The physical model shown in Figure 2.20 is organized around the largest scale pieces of the system — the subsystems. In large systems, independent teams usually develop subsystems; therefore, it makes sense to maintain this distinction in the model. Subsystems are constructed primarily of instances of classes from multiple domains. Put another way, each subsystem contains (by composition) objects instantiated from different domains in the system.

The last major package is builds. This area is decomposed into subpackages, one per prototype. This allows easy management of the different incremental prototypes. Also included in this area are the test fixtures, test plans, procedures, and so on, used to test each specific build for both the integration and validation testing of that prototype.

Figure 2.20   A subsystem-based model organization.

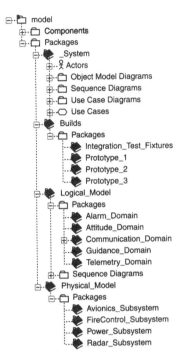

The advantage of this model organization is that it scales up to large systems nicely because it can be used recursively to as many levels of abstraction as necessary. The separation of the logical and physical models means that the classes in the domains may be reused in many different deployments, while the use of the physical model area allows the decomposition of system use cases to smaller subsystem-level uses cases and interface specifications.

The primary disadvantage of this model organization is that the difference between the logical and physical models seems tenuous for some developers. The model organization may be overly complex when reuse is not a major concern for the system. Also, many of the subsystems often depend heavily (although never entirely) on a single domain, and this model organization requires the subsystem team to own two different pieces of the model. For example, guidance and navigation is a domain rich with classes, but it is usually also one or more subsystems.

Physical Models

The last model organization is similar to the previous one, but blurs the distinction between the logical and physical models. This can be appropriate when most or all subsystems are each dominated by a single domain. In this case, the domain package is decomposed into one package for each domain, and each of these is further decomposed into the subsystems that it dominates. For common classes, such as bus communication and other infrastructure classes, a separate shared framework package is provided to organize them.

We've looked at some of the concerns and issues that drive model organization: application and team size, importance of reuse and other business culture concerns, and developer work flow. And we've looked at four different ways to organize your models. While each of these has been used successfully on appropriate projects, they each have pros and cons. Before you simply accept these model organizations as they are, evaluate your own business culture, work flows and project characteristics to see which, if any, of these existing schemes best fits your project — or how it might be modified to make your environment and product vision even better.

2.1.4 "Getting Started with Patterns"

by Luke Hohmann

Software patterns, which have their roots in the groundbreaking architectural design work conducted by Christopher Alexander, have helped software engineers drastically cut development time and improve software quality. Patterns accomplish this goal by capturing and sharing best practices in a compact, accessible, and extendible form. Patterns aren't fragments of well-documented source code (although they may contain source code) or a new API to an operating system. Instead, patterns are a highly structured literary form that documents a problem, the context(s) in which it may occur, and a good solution for resolving the forces presented by the problem. In the end, your system becomes easier to maintain and is of higher quality. Patterns document the wisdom of other developers' proven solutions in such a way that you can build on this wisdom through your own skill and experience. It is this sharing of wisdom that makes patterns so incredibly powerful.

Benefits of Patterns

Suppose, for example, you are building a system with a GUI and you want to provide for the undo, redo, and logging of commands (for debugging or macro support purposes). One way to solve this is to engage in a traditional, object-oriented analysis and design of the problem and then implement your design. In the process, you'll probably find you need to answer these important questions:

- How should commands be represented?
- Who should be responsible for creating, managing, and deleting commands?
- How do you distinguish between commands that can be undone and those that can't?

Unfortunately, resolving each of these issues chews up that absolutely critical development resource: time.

Another way to solve this problem is by building on the experience of other designers and using the Command Processor pattern, which describes how to separate service requests (commands) from their management and execution. You can find information about the Command Processor pattern in *Pattern Languages of Program Design, 2* by John Vlissides et al. (Addison-Wesley, 1996). This pattern not only describes the answers to the above questions (commands should be represented as objects; a CommandProcessor class manages commands, including deleting them when they are no longer needed; a state machine can distinguish between commands that can be undone from those that can't), it also highlights several other potential design and implementation "gotchas" (use two stacks to implement unlimited undo and redo, use the Composite pattern to implement macro capabilities, and so forth).

Note that the Command Processor pattern is not a piece of code wrapped inside a library that you simply link into your application. Instead, it is a pattern that describes, at a much more abstract level, how to solve the problem. Using it requires you to tailor it to meet the specific needs of your problem. A pattern provides guidance on how to create a solution; actually creating it is in your hands. In this way, a pattern can be used thousands of times, with each idiosyncratic implementation true to the pattern but distinctly unique. The uniqueness is a result of the context created by your problem.

A second key benefit of patterns concerns the broader and deeper issues by which we design systems. Most system design work can be characterized as trying to manage complexity through commonality and variability analysis. Relational databases, for example, rely on the rules of normalization to determine what is common and what is variable, while object-oriented analysis and design is based on the creation of specific taxonomies of classes related through aggregation and inheritance. But what happens when this kind of commonality and variability analysis breaks down, such as when you want to structure the interaction between a group of objects in a meaningful way? Patterns provide a means of capturing and sharing the interactions between objects in ways that go beyond commonality and variability analysis.

A third benefit of patterns is their applicability to many different problem domains. Because patterns concentrate on proven solutions to recurring problems in specific contexts, they can (and have) been used to describe many different kinds of problems. A sample of pattern languages and their applicable domains include:

- *Caterpillar's Fate*, which describes the processes you should follow when developing concurrent systems
- *Crossing Chasms*, which describes how to integrate object-based systems with relational databases (a topic covered in detail in The Unified Process Construction Phase, volume 3 in this series)
- *A Generative Development-Process Pattern Language*, which provides guidance for structuring projects based on studies of successful software projects
- *Reactor*, which describes how to build concurrent systems.

Patterns aren't just a single-point technology solution, like a different way to organize a database for more efficient analytical processing. Instead, they are a literary form designed to convey wisdom. This form has surprising versatility.

A final benefit of patterns is the way they create a shared language for reasoning about solutions to problems. Consider two designers skilled in the use of design patterns. Because each shares the same relative understanding of the solution, they can discuss alternatives at the higher-level abstractions provided by patterns rather than talking at a detailed level about two different design decisions.

What's in a Pattern?

In a way, patterns are an extremely sophisticated form of documentation. Although their specific structures may vary, patterns convey the following information. Certain literary forms that are used to structure patterns make these sections explicit.

Name. A name that captures the meaning or intent of the pattern in a way that facilitates communication.

Intent. A single sentence that summarizes what the pattern does, describing the design issue or problem it addresses.

Problem. A description of the problem the pattern solves. Note that two different patterns may have a similar intent, but solve slightly different variations of the same problem. This intent is often founded on the deeper principles that guide the creation of well-crafted software (low coupling, high cohesion, encapsulation, and so forth).

Context. The context helps you compare the specific forces that characterize your problem to the pattern's intent so you can make a wise decision on the pattern's use. For example, the Flyweight pattern uses sharing to support large numbers of fine-grained objects. However, its applicability is based on forces that can only be found in the context of the application. For example, you can't use Flyweight if each object must maintain a unique identity or has a unique state that cannot be made extrinsic.

Forces. Forces guide you in the specific implementation of the pattern. I like to think of them as a means of ensuring the solution I'm about to implement creates a foundation that not only reduces entropy now, but works to keep it low as the system evolves over time.

Solution. Ultimately, patterns must deliver a solution. The description usually addresses the broadest possible context, but uses enough detail so you can solve the problem directly. Patterns that deal with coding issues, for example, will often include code fragments to help ensure the meaning of the solution makes sense.

Sample. Many patterns include a sample of how the pattern was previously used to solve a specific problem.

Consequences. Also known as the resulting context, this section concludes the pattern by describing the expected state of the system after the pattern is applied. In a pattern language, the resulting context leads you through the successive application of different patterns to engineer a robust solution that effectively addresses all of the problem's complexities (and the new problems introduced by a specific solution).

Good pattern writers balance the amount of information conveyed with the length of the document. More specifically, a good pattern ranges from one to four pages in length; brevity is preferred. Like yourself, pattern writers are often busy developers. Writing a novel isn't necessary to convey the information needed to describe and solve a problem.

Another important rule is the rule of three: unless the solution described in the pattern has been used in at least three systems, it is rarely considered a pattern. That doesn't mean novel solutions aren't worthy of precise documentation or sharing; they may represent the heart of an important but specialized business algorithm. However, such solutions don't adhere to the natural meaning of the word "pattern." Thus, many pattern writers introduce (or close) a pattern by describing the systems that have implemented the pattern successfully. Skeptical developers can use these descriptions to judge the benefits of using a particular pattern in their work.

Identifying and Using Patterns

There are many ways to become familiar with patterns and start using them. You can browse through web sites and peruse the online articles, or you can purchase a book on patterns. But reading about patterns won't necessarily provide the skill and experience you need to apply them. Successful use of patterns requires a bit more effort.

First, you must clearly understand the problem you are trying to solve. I've found that stating the problem in a concrete manner, using one or two sentences, helps a great deal. You should also have a broad understanding of the context in which the problem is situated and the forces in the system that will influence your decision. Are you more interested in maintainability or efficiency? Do you have three developers or seven? Is the context stable or likely to change?

Second, you must ask yourself if others may have faced this (or a similar) problem in the past and if you are willing to search for and apply one of their solutions. Unfortunately, for numerous reasons, this step is harder than it sounds. Developers are often wary of solutions developed by others (the classic "not invented here" response). Developers also face severe time constraints and searching for the right pattern may not appear to be a luxury they can afford. But, if you are facing a problem that you think may have been solved before, you certainly owe it to yourself (and your fellow developers) to look for a pattern-based solution.

The third key element is the ability to put the specific context of your problem in abstract terms. I've found that restating a problem in an abstract manner helps people move from specific requirements to the more abstract realm of patterns. One way of doing this is to change the names of the classes in your concrete problem to more generic names that are reflective of an abstract problem. Another is to imagine you are trying to solve this problem for a slightly different system. A third approach is to think of the problem only in terms of abstract interfaces.

To illustrate, suppose you must write some glue code to connect a C++ system to a set of FORTRAN libraries. The specific problem might be creating a C++ object for Susan and Ramir that helps them use the FORTRAN SIGNLLIB library for signal processing. The abstract problem might be providing a means of adapting one system to use a set of features and functions provided by another. Another abstract approach would be to ask the question, "How can I create a subsystem that provides a simple and cohesive interface to another, more complex subsystem?"

If you are familiar with the patterns described in the book *Design Patterns: Elements of Reusable Object-Oriented Software* by Erich Gamma et al. (Addison-Wesley, 1994), you've probably identified two good candidates for solving this problem: Adaptor, which provides guidance on how to adapt one interface for use by another; and Façade, which provides guidance on creating simplified interfaces to complex subsystems. The specific pattern you choose would depend on the forces presented by the original problem. For example, Adaptor is more appropriate when you want to simply adapt object interfaces; Façade is a better choice when part of the design goal is to simplify the interface to the subsystem. Suppose the FORTRAN library required fairly elaborate initialization procedures, and you wanted to hide these details from your fellow developers. In this case, you would probably choose Façade, implementing it in such a way as to minimize initialization complexity.

Forming and asking abstract questions is an important step in using patterns effectively. Different forms of the same abstract question may yield slightly different results. Thus, the best abstract questions consider the surrounding context of the original problem. Over time, you will become more effective at forming abstract questions. Once you have identified the right pattern, you will find the combination of your specific problem statement and the pattern itself will help guide you to a good solution.

There is no guarantee you'll find a pattern that solves your problem. This isn't a failing of patterns, but instead the realization that it doesn't make sense to capture every solution in pattern form. Remember, patterns are designed to capture the proven solutions to commonly recurring design problems, not to provide a solution for every design problem. Sometimes you must resort to more general principles of system design to solve your problem (such as information hiding, loosely coupled and highly cohesive components, stable interfaces, and so forth).

Patterns Aren't a Silver Bullet

Although I am an enthusiastic supporter of patterns, they aren't a silver bullet that can solve every problem in software development. Fundamentally, great software is written by adhering first and foremost to the timeless principles of software design: information hiding, encapsulation, well-defined and semantically meaningful interfaces, source code with variables and formatting chosen with care, and so forth. Patterns are not a replacement for knowing how to apply these principles as a professional software developer.

Coupling, for example, is a measurement of the degree of interconnectedness between two components. It isn't a pattern. And principles such as "low coupling is good" aren't patterns. But the techniques you use to achieve low coupling can be captured as patterns. By capturing these techniques as patterns, you can define the design decisions that must be made in realizing a specific lower coupling. (Sometimes the right design decision is to increase coupling; as always, the best decision depends on the specific context of the problem you are trying to solve.)

Another problem is pattern explosion. As more and more developers discover patterns and begin to use the pattern form to document and share their knowledge, the number of published patterns continues to grow. Too many patterns may be overwhelming. Choosing among the 23 different patterns in Design Patterns can be difficult. How quickly and effectively could you choose between 230? I believe that, over time, we will continue to see the development of domain-specific pattern languages that provide a level of organization over collections of patterns. In this manner, the pattern language will become your entry point into choosing the right patterns to address your specific problems.

A deeper problem exists in the misuse of patterns. I've witnessed developers striving to force-fit patterns to their problems. Patterns are most effective when they feel as if they resolve your problem in a relatively seamless manner. The overall entropy of the system decreases and a feeling of greater calmness occurs in the resulting implementation.

One perceived problem with using patterns in C++ is that solutions based on patterns usually contain more classes than those that don't. This is a direct result of enhanced encapsulation and the use of abstract base classes to specify stable interfaces. Unfortunately, many developers consider adding extra classes in C++ painful. My response? Get over it. If a design pattern is the right way to solve the problem, use it. (To be fair, however, adding a class in C++ is a bit of a pain: you have to create a header file and an implementation file, add it to your make file, and so forth.) The time it takes to do this in support of a pattern will be won back many times over through enhanced maintainability and comprehension.

A final problem with patterns concerns the larger community of developers in which you work. Are you the only member of your team who knows or uses design patterns? If this is the case, patterns may actually seem to detract from teamwork. For example, you may find fellow developers mystified by your solutions and that you haven't yet found the advantages of shared design vocabulary. This is a real problem. What makes patterns so powerful also makes them inaccessible for the uninitiated. If you find yourself in this situation, I encourage you to organize a series of brown bag lunches so your team can discuss how to use patterns.

Sharing the Wealth

Since today's development teams are often asked to bring products to market faster, with higher quality and enhanced long-term maintenance characteristics, I've become skeptical about the claims associated with new development technologies. Quite simply, many new technologies fail to deliver what they promise.

I've found patterns represent a significant breakthrough in improving software development practices. They do this by enhancing the most powerful tool that exists in problem solving: your mind. Specifically, patterns don't promise to automate the creation of source code from specifications or diagrams. They aren't a form of component-level reuse. Instead, they represent the wisdom of other developers captured in a literary form that is designed to be shared efficiently and effectively.

In my company, we use numerous patterns in the development of our software systems. For example, of the 23 patterns described in *Design Patterns*, we've used five of the six creational patterns, four of the seven structural patterns, and nine of the eleven behavioral patterns. We've also used several of the patterns in Craig Larman's GRASP pattern language, as well as several patterns from the *Pattern Languages of Program Design* book series (Addison-Wesley, 1995, 1996, 1997, and 1999). I'm confident we've saved thousands of dollars and have dramatically improved the quality of our source code by building on the proven solutions of others. I'm equally confident you'll find similar value in your own development projects through the use and adoption of patterns.

2.1.5 "CRC Cards for Analysis"

by Nancy Wilkinson

When building an object-oriented model of an application, the focus of the early CRC card sessions is on analyzing and describing the problem. During this problem-modeling phase, the group identifies entities that are relevant to the application domain. These entities become analysis classes, and the behavior necessary to describe the activity of the system is represented as responsibilities of these classes. Any system activity is the activity of an object, or set of objects, of these classes.

In this article, we emphasize the conceptual/problem-modeling or analysis aspects of CRC cards and the CRC card process. We then discuss various topics of interest during this early phase of the project. We investigate how CRC cards can be used by people who want to move their projects to object-oriented development but have the constraints of legacy systems. We then discuss reuse considerations in problem modeling.

Analysis vs. Design

The line where analysis ends and design begins in the lifetime of an application may be somewhat fuzzy. This is true in every project, but especially in small- to medium-sized projects of, say, two to ten team members. In these small-team efforts, the design model may be built directly with no separately produced problem model. Nevertheless, the distinction between which part of the model describes the problem and which describes the solution is an important one.

First of all, the analysis and design represent different views of the system; thus, they form the vocabulary for different groups of people involved in the project. For example, the analysis will have high client visibility. If you need to interact with customers about what the system will do, a separate problem model is important. Classes critical to the design and implementation phases, such as queues and database management interface classes, are meaningless to users. Also, the makeup of the group in CRC card sessions shifts as it moves from analysis to design. Early sessions that concentrate on problem modeling require more domain expertise than design expertise. As the group moves to solution modeling, the emphasis shifts to design expertise.

Even if no separate problem model is produced, the probability of building the right product will be greater if you focus early on what the application does and make your design decisions within that framework. For this discussion, we will assume that a stable problem model is reached before the design stage is entered at full steam.

Strengths of CRC Cards for Analysis

CRC cards have particular strengths at each phase of development. The advantages of using the cards for problem modeling are as follows:

Common Project Vocabulary.   People from all phases of the project can be involved in CRC card sessions for problem modeling from the outset. Hence, the names of classes and responsibilities are seen, heard, and understood by everyone involved in the project. Most important, these terms are created, debated, and agreed upon by project members at the start of the process. This leads to a common product and project vocabulary across development roles and enhances the concept of a common team goal.

Spreading Domain Knowledge.   As we have stated previously, the good news and bad news about objects is that domain entities participate directly in the application, and good or bad domain knowledge has a high impact on the quality of the application. This implies that it is vital for the people building the application to have a good working knowledge of the domain. CRC card sessions get the right people in the room and facilitate discussion. Domain knowledge is spread in a very real way in the sessions because participants actually experience the workings of the system. This serves to teach and clarify in a way that reading requirements or looking at a set of pictures cannot.

Making the Paradigm Shift.   The role-playing of the system model also facilitates the spread of knowledge about the object-oriented approach to problem modeling. The paradigm shift is internalized by participants through the experimental and experiential nature of the sessions. Every participant is an object, and everything that happens does so because an object makes it happen.

Live Prototyping.   At this stage, people who are not participants in the modeling of the system can be shown the model by running through scenarios. This can help give designers, developers, and testers a basic knowledge of the application or bring new project personnel up to speed. This technique can also be used to show customers the application. In a sense, the session can serve as a live prototype of what the system is supposed to do. And it is done in a setting in which feedback from the customers is encouraged and can be easily incorporated.

Identifying Holes in Requirements.   A CRC card session provides an opportunity to literally walk through the requirements and identify holes or ambiguities. It is not uncommon during the simulation of a scenario to have someone ask, "What needs to be done next to fulfill this responsibility?" If the information in the requirements is vague, ambiguous, or missing, a domain expert needs to clarify the point. If no domain expert knows the answer, someone can commit to meet with the customer to get the answer. In either case, the author of the requirements sees the problem with the document and what needs to be done to correct it.

The Library Application Problem

Participants in the sample session presented in this article will be modeling a library system for checking out books, returning books, and searching for books. The requirements for the system are as follows: This application will support the operations of a technical library for a research and development organization. This includes the searching for and lending of technical library materials, including books, videos, and technical journals. Users will enter their company IDs to use the system; and they will enter material ID numbers when checking out and returning items.

Each borrower can be lent up to five items. Each type of library item can be lent for a different period of time (books four weeks, journals two weeks, videos one week). If returned after their due date, the library user's organization will be charged a fine, based on the type of item (books $1 per day, journals $3 per day, videos $5 per day). Materials will be lent to employees with no overdue lendables, fewer than five articles out, and total fines less than $100.

CRC Elements at the Analysis Stage

The following discussion describes the types of classes, responsibilities, and scenarios that are unique to the problem-modeling phase of development. These are contrasted with the same CRC elements involved in design.

Classes.   The classes identified in problem-modeling sessions directly reflect concepts and entities that are part of the domain being modeled. They tend to be the concepts and terms that domain experts use to describe the problem and what they want the application to do. Other classes, such as representations of implementation-level abstractions or interfaces to system resources, will be added as the system is designed. Classes at this stage are those that describe what a system does. For example, the classes we identified for the Library domain are application-level or analysis classes.

The classes in problem modeling may include things that are outside the software system but useful in describing the way an application should behave. For example, in the Library model, the User may have been useful when describing what the system was to do but will not be useful as a software abstraction for solving the problem. Other classes in the problem model may serve as a place holder for functionality, which will be performed by more than one design class.

A CRC card session provides an opportunity to walk through the requirements to identify holes or ambiguities and what needs to be done to correct them.

At the problem-modeling stage, the distinction between classes and objects may not be completely clear. Some object of the class has responsibility for doing something, but exactly which object does it and how is not completely specified. Also, dynamic issues, such as who creates an object and the lifetime of the object, are usually not dealt with until the design phase. Classes may also be relegated to subsystems at this stage of development. This enables parallel development from analysis through prototyping.

Scenarios and Responsibilities. Scenarios are detailed examples of the functioning of the system. They are the answers to questions that begin with "what happens when..." At this initial stage of development, only scenarios that explore what the system is supposed to do are considered. For example, in modeling the Library application, an important functionality of the system is to be able to check out items. Scenarios for this function included such questions as "What happens when Barbara Stewart, with one outstanding book not overdue, checks out a book, *Effective C++ Strategies*?" Most of the scenarios at this stage are driven by user-level views of the system.

Accordingly, the responsibilities assigned to classes during the problem-modeling CRC card sessions are of a high-level nature. Some may appear explicitly in the requirements documentation for a system. Others are derived by the group when executing the high-level scenarios. They are the steps that must take place to make the scenario happen. But they still correspond to what happens rather than to how it happens. For example, the sets of responsibilities that describe the fulfillment of the previous scenario might be the following:

- User asks Librarian to check out a book
- Borrower checks its status
- Book checks itself out
- Borrower updates its set of Lendables.

But how does a Book object check itself out? You may decide that it updates its borrower, in or out status, and due date knowledge. Are these steps part of analysis or design? This is where things get fuzzy. Let us assume for the time being that we consider these steps to be part of the problem model. There are still lower-level considerations (we assume this second-generation system will be very much like the existing application with some additional features or functionality). The existence of such a legacy system will undoubtedly have an impact on the development of the second-generation product. What are the choices open to these projects, and how can CRC cards support them?

Some projects jump in with both feet, rewriting the application from scratch. A more gradual approach to object-oriented development, sometimes used by projects moving from C to C++, relies on reusing some or all of the existing system. These decisions affect the project at every stage and need to be incorporated at the start of the sessions in the analysis phase. Here are a few ways in which CRC cards can be used to support these projects.

The Full-Scale Approach. If a group is considering moving to an object-oriented approach, the best business decision it can make for the long term is to write the new application from scratch using object-oriented techniques. This is especially true if the existing system is old or poses maintenance problems in its present form. Such a project, if the costs of training and learning are built into the schedule and the budget, can convert a group of people to object-oriented techniques while gaining the benefits advertised by the object-oriented approach. It has been suggested that over time, the costs of not taking this full-scale approach are greater than these up-front costs.

CRC cards support this rewriting of a system much as they do the building of a new application, usually by partitioning the system and remodeling one subsystem at a time. But groups with existing systems bring additional domain knowledge into the sessions. The set of requirements for the new application drive the sessions, but are supplemented by knowledge about the existing system. Such a group also understands the strengths and weaknesses of the existing implementation, which can be an advantage as it moves into design. This implementation-level acquaintance with the system, however, can be a disadvantage in the problem modeling of the new application because participants must separate and describe what the system does without being biased by how it currently does it.

One Subsystem at a Time. Rewriting second-generation applications using object-oriented techniques will be costly in the short term. And if the existing system is of high quality and has a healthy place in the market, abandoning it is unnecessary. However, if a portion of the system is separable from the rest, that portion can be implemented as a separate subsystem using object-oriented techniques. This may be an existing subsystem, such as the user interface, or one built to perform the new functionality required by the second-generation application.

The disadvantage of this choice is the extra work involved in bridging subsystems written using different approaches. Also, the benefits of the object-oriented approach will only apply to the one new feature. On the other hand, this hybrid approach may require training only a subset of the people on the project. It certainly allows projects to gain experience with object-oriented techniques without dedicating all the time and resources up front.

CRC cards are a popular choice to guide the analysis and design of one subsystem while training team members in the object-oriented approach. Even projects that normally use formal methodologies because of the size of their applications enjoy the informality of CRC cards for one portion of the system. System requirements that apply to the subsystem being built or rebuilt drive the scenarios. In addition to the cards to model the subsystem under consideration, the session will include a class or set of classes to hide the details of the interactions with other parts of the application. We call these classes interface classes.

For example, if a group wants to add a Report feature to an existing application, there may well be a Report class whose responsibilities represent the overall services that the new subsystem proimplicit in these steps that are unquestioningly part of design.

What does it mean for a Book to update its in or out status? Is there an attribute or data member that represents this state information that can be changed? Or does it make a change to a database table somewhere on the fly? How about Borrower's ability to borrow? These questions are not answered at this stage, but are deferred to design.

Later we will see how scenarios are used in CRC card design sessions to further refine the responsibilities, even to the level of member functions and their implementation. But at this stage, potentially many methods or class member function operations are collectively represented in a single piece of English text.

When Do You Have a Stable Model?

As more and more scenarios are executed in a CRC card session, the cards on the table fill out and the problem model begins to stabilize. The first sign that the model is near completion is that the execution of new scenarios, even exceptional scenarios, does not cause new classes or new responsibilities to be added to the model. Nor does executing new scenarios cause responsibilities to change classes to better handle a new situation.

The real test that a group is reaching a stable model is when a subtle shift occurs in the way the model can be used. At some point the model begins to provide information about the problem domain rather than require it. It becomes a place where people are getting, rather than putting, their domain knowledge. In essence, it becomes the source of information rather than a sink for people's perceptions.

The model has become a framework for the design stage, where more information about how the system should work can be built on top of a working model of what the system does.

Library Application Problem Model

One product of problem modeling with CRC cards can be that the classes and responsibilities are at a high level, corresponding to what the system is supposed to do, not how it does it. Figure 2.21 provides a picture that represents the Library application problem model. This picture, plus a set of scenarios, can be used to describe what the application will do.

CRC Cards and Second-Generation Systems

Thus far we have presented CRC cards as a technique to build an entirely new application using an object-oriented approach. But many organizations that move to object-oriented development have an existing/legacy base of code and want to construct a second-generation system vides, such as generating reports. This class will collaborate with Report subsystem classes to carry out these responsibilities.

In the implementation, the existing system is updated to call functions of this class to invoke the new feature. There may be other interface classes, for example, Database or User Interface, whose responsibilities are the services of these subsystems that are required by the Report subsystem (gather data or display). The eventual implementation of these responsibilities is in terms of the code that already exists to perform these actions.

Reengineering Existing Systems. Some projects are dedicated to producing an object-based or object-oriented second-generation application but want to reuse much of the existing code. We will refer to this approach as reengineering existing systems. It is probably the least common approach of the three we mention.

Figure 2.21   Library application problem model.

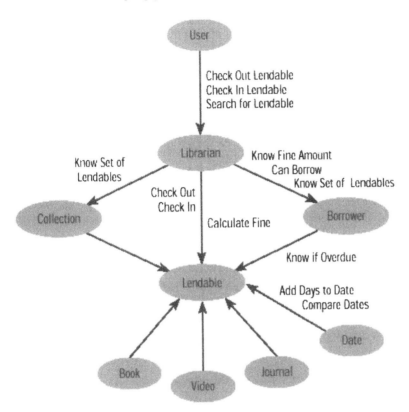

This approach is much like the full-scale approach, except that the final model will necessarily be constrained by the existing system's implementation, at least at a low level. The advantages to this approach are that the resulting system will be implemented with objects, and the entire team begins to move to the new paradigm. The disadvantages are that the design, because of the constraints of the structured code, may not realize the full benefits of an object-oriented approach.

CRC card sessions to reengineer systems should build an initial object-oriented problem model and high-level design just as if the system were being rewritten from scratch. This ensures that team members gain experience in pure object-oriented analysis and design without the intrusion of the existing code.

Some analysis of the existing system should then be done. The data structures and the functions that operate upon them should be identified and grouped and CRC cards representing the resulting abstractions produced. The responsibilities on these cards are the functions that exist to operate on the enclosed data structure. These cards can then be used in the second-pass CRC card design sessions.

For example, say that in the C version of a work management system a structure called WorkTicket is central. Separate functions exist to operate on the Ticket, such as assignTicket() and close Ticket(). In sessions to reengineer such a system, a card is created to represent the Work Ticket abstraction, and it is assigned initial responsibilities, including assign and close. Later, each card derived from existing abstractions is implemented as a wrapper class. A wrapper class is one that encapsulates the existing data structure and implements its member functions as simple calls to the functions that operate on this structure.

Creating Classes for the Library Model

The first step in modeling an object-oriented system, and thus the first step in the CRC card process, is finding classes in the problem domain. Classes can be listed with the entire group in a first step before creating any cards. Alternatively, classes can be encountered and created as part of the execution of the scenarios. Different groups may find different methods useful.

Starting a group off by brainstorming classes together emphasizes that this is a group effort, and it gives everyone a chance to get involved in the process from the beginning. It seems to be a nonthreatening way to introduce a group to a new situation and new technology. In addition, it provides a framework of classes, however preliminary (a house of cards?), upon which to begin distributing responsibilities during the walk-through of the scenarios. New classes will certainly be discovered during scenario execution, but the major classes can usually be identified at this early stage.

In the Library model, the classes and their descriptions emerge as the following:

Librarian   The object in the system that fulfills User requests to check out, check in, and search for library materials.

User   The human being that comes to use the library.

Borrower   The set of objects that represent Users who borrow items from the library.

Date   The set of objects that represent dates in the system.

Lendable   The set of objects that represent items to be borrowed from the library.

Book   The set of objects that represent books to be borrowed from the library.

Journal   The set of objects that represent technical journals to be borrowed from the library.

Video   The set of objects that represent videotapes to be borrowed from the library.

Reuse and CRC Cards in Problem Modeling

Reuse is not a concept unique to object-oriented development. Reuse has been a topic of exploration for years (see the article "Reusability Comes of Age," by Bill Tracz in *IEEE Software*, Vol. 7, No. 6, 1987).

But object-oriented systems seem ideally suited to capitalize on the phenomenon. Reuse of general-purpose and special-purpose classes and frameworks is becoming widespread. Reuse of analysis classes has not happened on a large scale. But reuse of locally constructed domain classes intended for use by a particular project or family of projects within an organization is a reality.

The problem-modeling phase is not too early to begin thinking about the potential for reuse in the application. Reuse of domain-level classes or architectural frameworks should be considered here. When entering a CRC card session, it is important to know what domain classes and frameworks exist to help model the problem. Then, during CRC card sessions, intelligent decisions can be made about whether the existing classes can be used for the new application. The Library application was the first such system to be written using object-oriented techniques; therefore, it could not take advantage of existing domain-level classes. But the next time a Library application is written, perhaps to support billing, purchasing, or administrative tasks, some of these cards, or sets of cards, could be reused in a problem model for the new application.

Pictures such as Figure 2.21 can be useful in identifying classes, and thus cards, in a problem model that may be reusable. Classes, or groups of classes, that have arrows only pointing into them are good candidates for reuse because they operate independently of the other classes in the model. For example, Librarian might be the worst candidate for a stand-alone reusable component from the Library model; however, the Lendable hierarchy is a potentially good candidate to be used in another application.

CRC card sessions can also help create new domain classes or generalize existing ones to make them more reusable across applications. Cards for these classes could be stored in a card catalog for the organization. This electronic catalog could be searched for appropriate cards or sets of cards (frameworks), and copies of the cards could then be generated for the new project's use. The classes may or may not have designs and implementations associated with them. The idea of reusable problem models and how best to store, search for, and generalize the cards that represent them is an interesting area of research.

3

Best Practices for the Requirements Workflow

Introduction

The purpose of the Requirements workflow is to engineer the requirements for your project. During the Inception phase, you need to identify the requirements for your software to provide sufficient information for scoping, estimating, and planning your project. To do this, you often need to identify your requirements to the "10–20% level" — the point where you understand at a high-level what your project needs to deliver, but may not fully understand the details. You are likely to develop several artifacts as part of your requirements model, including, but not limited, to an essential use-case model (Constantine & Lockwood, 1999; Ambler, 2001), an essential user interface prototype (Constantine & Lockwood, 1999; Ambler, 2001), a user-interface flow diagram (Ambler, 2001), and a supplementary specification (Kruchten, 2000). A supplementary specification is a "catch-all" artifact where you document business rules, constraints, and non-functional requirements. During the Elaboration phase (Ambler & Constantine, 2000a), you will evolve your requirements model to the 80% level and will finalize it during the Construction phase (Ambler & Constantine, 2000b).

Your goal during the Inception phase is to develop your requirements model to the "10–20%" level.

There is much more to the Requirements workflow than simply developing a requirements model. You must also elicit the needs of your project stakeholders so that the overall vision for the project may be negotiated. You should strive to define a common vocabulary by developing a project glossary — an artifact that should be started during the Inception phase and evolved throughout the entire project lifecycle. This glossary should include a consistent and defined collection of both business and technical terms applicable to your system. Furthermore, you should define the boundary and scope of your system — the goal being to identify the requirements that your project team intends to fulfill as well as those it will not. Finally, during the Inception phase, you will develop a vision statement for your project, summarizing the high-level requirements and constraints.

Developing a requirements model is just one part of the Requirements workflow.

What are best practices for the Requirements workflow? First, involve the real experts. When defining requirements, you should strive to get the people who best know the problem domain involved. This very often means getting front-line workers, not just their managers, involved in the process. Second, document the source of each requirement. The source is often just as important as the requirement itself, as it helps to verify the credibility of the requirement. Furthermore, it's information that can be used to expand on the requirement during your Business Modeling workflow (Chapter 2) and Analysis & Design workflow efforts. Third, realize that the user requirements definition is a start to modeling. The work that you do as part of the requirements workflow will directly affect how much work will be needed to model the application. The more detailed and thorough the requirements are now, the less work will be needed later during modeling. Fourth, perform requirements triage early. Recognize the fact that you aren't going to be able to deliver every single requirement requested by your users; therefore, you need to prioritize and deliver the most important requirements that you can identify. Fifth, have a technical writer aid in the documentation process, a significant portion of requirements definition. Because technical writers specialize in documentation, and because they are often paid significantly less than requirements modelers, it makes sense to have them document requirements, allowing your modelers to concentrate on what they are good at — identifying requirements. The act of including a technical writer on your team is often referred to as the *Mercenary Analyst* (Coplien, 1995) organizational pattern.

Perceive that which you cannot see. — Sensei Rick Willemsen

An important best practice for this workflow is to recognize that there are many ways to engineer the requirements for your system, as you can see depicted by the solution for the *Define and Validate Initial Requirements* process pattern (Ambler, 1998b) of Figure 3.1. An important point to note is that there are many requirements engineering techniques, that use-case modeling is just one of many. Use-case modeling an excellent mechanism to identify and understand the behavioral requirements for your software, but it is not the only technique that you have available to you. Use cases are used as key input into your planning process — an activity of the Project Management workflow. An important best practice is to prioritize your use cases by both technical and business risk so that you can address the riskiest use cases in the earliest iterations of your project, reducing your project's overall risk early in the lifecycle.

Use cases are the tip of the requirements iceberg.

Figure 3.1 The Define and Validate Initial Requirements process pattern.

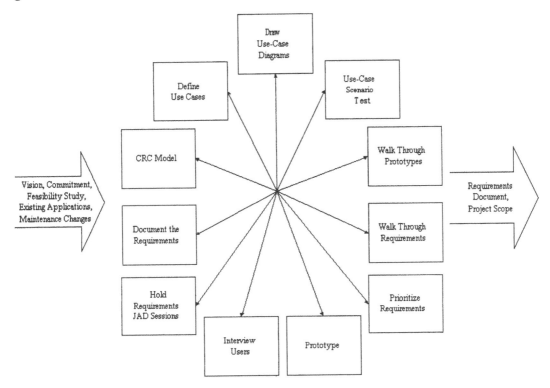

Another success factor for modern development is the recognition that modeling, including requirements engineering, is an iterative process. As you can see in Figure 3.1, requirements engineering is comprised of techniques for identifying, documenting, and validating requirements that can and should be applied iteratively. With an iterative approach, you typically apply the various techniques in any order that you need to, working on a portion of one artifact, then switching over to another. In fact, working on several deliverables simultaneously is a best practice that your organization should consider adopting. It is quite common to find requirements engineers working on a use case model to document behavioral requirements, a Class Responsibility Collaborator (CRC) model to document the key domain entities, and a prototype to understand the requirements for the user interface at the same time. Although this appears to be time consuming on the surface, it actually proves in practice to be very efficient. Because no one artifact is sufficient for all of your requirements engineering needs, you will find that by focusing on a single artifact, you will quickly become stuck when you don't know how to use it to document something or its scope becomes too narrow to fully investigate the problem at hand. Instead, it is more effective to work your use case model for a while and once you run into problems, see if you can address them by hand-drawing some screens or reports. Then, when you run into roadblocks, you can work your CRC model for a bit or move back to your use case model.

The Requirements workflow focuses on the identification and documentation of requirements, whereas validation techniques, such as walkthroughs and inspections, are activities of the Test workflow (Chapter 4). In fact, you will see that these two work flows are highly dependent — in "A Business Case for QA and Testing" (February 1999, Section 4.4.1), Nicole Bianco argues that improved requirements decreases the chance of introducing analysis errors and enables you to identify and develop more accurate tests. In "Reduce Development Costs with Use-Case Scenario Testing" (July 1996, Section 4.4.4), Scott Ambler shows how testing early in the lifecycle improves the quality of your requirements models.

Working several models in parallel is significantly more productive than focusing on one model at a time.

Another best practice for the Requirements workflow, and for software development in general, is to document your assumptions as you go along. There are several benefits in doing this. First, by documenting your assumptions, others can put your work into context and will have a greater understanding of what it is you are trying to achieve. Second, you will now have a list of issues that you need to look into — to investigate and resolve any assumptions as early as possible. Assumptions are dangerous because you are basing decisions on simple guesses.

Document your assumptions.

Requirements engineering is a fundamental part of virtually every software development task. If you are working on your organization's business/domain architecture, you should base your efforts on the high-level business requirements captured in your enterprise model. If you are working on your organization's technical architecture, you should base your efforts on the technical requirements for your organization, such as the need to support distributed computing in a seven days a week, twenty-four hours a day (24/7) manner. If you are evaluating tools, standards, or guidelines for your project team, you should first start by defining the requirements for the items (i.e., have a shopping list). If you are designing your system, even if it is highly technical software such as an operating system or a persistence layer, you should start by defining the requirements for what you are building. There is a name for developing software without requirements — hacking.

Without defined and accepted requirements, you have nothing to build.

An important best practice — or perhaps general philosophy would be a better term — is accepting the fact that change will occur. Requirements change over time for a variety of reasons. The business environment may change, perhaps due to new government legislation or the actions of a competitor (or maybe even business partner). Your project stakeholders may change their minds, perhaps because they weren't certain of the requirements to begin with or because the work of your project team has caused them to rethink what they were asking for. Your project stakeholders might even change, perhaps a new senior manager becomes responsible for your project part half way through its development, bringing with them a new vision. Furthermore, because it is impossible to precisely specify the requirements for a system, you must expect that your own understanding of the requirements will change over time. (In this case, the actual requirements haven't changed, just your understanding of them.)

Change happens.

Finally, you must understand that the requirements for your project must reflect, and be a detailed subset of, the high-level requirements for your enterprise. Although few organizations have a well-documented enterprise requirements model — or even a poorly documented one for that matter — the fact still remains that the system your team is building, or buying, must fit into the overall organizational environment. (Understanding the organizational environment is part of the Business Modeling workflow, Chapter 2.) If such a model exists, start by reviewing your enterprise requirements model and carve out the portion of it that you are working on — effectively using it as a basis for your own detailed requirements model. Enterprise requirements modeling is an aspect of the Infrastructure Management workflow, a topic covered in detail in *The Unified Process Elaboration Phase* (Ambler & Constantine, 2000a) and *The Unified Process Construction Phase* (Ambler & Constantine, 2000b).

Start at your enterprise requirements model.

3.1 Putting the Requirements Workflow into Perspective

Ellen Gottesdiener, in "Decoding Business Needs" (December 1999, Section 3.5.1), puts the entire requirements workflow — in fact, the entire Inception phase — into perspective. She shows how facilitated workshops, such as Joint Application Development (JAD) sessions (see Section 3.2), reduce the risk of scope creep from 80% to 10%, accelerate the delivery of early lifecycle phases by 30%–40%, and provide a 5%–15% overall savings in time and effort throughout the entire lifecycle. Not bad. Gottesdiener also shows three practical ways to involve business customers in the software development process: (1) establishing a project charter, (2) using facilitated workshops, and (3) addressing organizational change. A *charter*, similar conceptually to the business case and vision documents that you develop as part of the Project Management workflow (Chapter 5), begins with the end in mind, defining the who, what, when, where, why, and how of the project. A *facilitated workshop* is a planned, collaborative event in which the participants, led by a neutral guide, deliver products in a concentrated period of time. Organizational change management (OCM) is, in fact, what software development efforts are designed to do because the business goals and objectives that drive development are based on the expectation of change. When a software project fails, it often has nothing to do with the technology and everything to do with people factors.

During the Inception phase, your team needs to identify, and then build, a rapport with its project stakeholders, including direct users of the system, front-line managers, senior managers, as well as operations and support staff. How should you interact with your project stakeholders? How should they interact with you? What expectations should each group have of the other? In "Customer Rights and Responsibilities" (December 1999, Section 3.5.2), Karl Wiegers, author of *Software Requirements* (1999), addresses these issues head-on. He believes that a shared understanding of the requirements approval process should alleviate the friction that can arise between development teams and project shareholders — software success depends on developing a collaborative partnership between the two groups. He points out that the stakeholder-developer relationship can unfortunately become strained or even adversarial. Problems arise partly because people don't share a clear understanding the requirements and who the customers are. To clarify key aspects of this partnership, he proposes both a Customer's Bill of Rights and a corresponding Bill of Responsibilities, summarized in Table 3.1.

Table 3.1   Project stakeholder's rights and responsibilities.

Rights	Responsibilities
• To expect analysts to speak their language • To have analysts learn about their business and objectives • To expect analysts to create a requirements model • To receive explanations of requirements work products • To expect developers to treat them with respect • To hear ideas and alternatives for requirements and their implementation • To describe characteristics that make the product easy to use • To be presented with opportunities to adjust requirements to permit reuse • To be give good-faith estimates of the costs of changes • To receive a system that meets their functional and quality needs	• To educate analysts about your business • To spend the time to provide and clarify requirements • To be specific and precise about requirements • To make timely decisions • To respect a developer's assessment of cost and feasibility • To set requirement priorities • To review requirements models and prototypes • To communicate changes promptly to the requirements • To follow your organization's requirements change process • To respect the requirements engineering processes that developers use

How do you go about the process of engineering requirements? Although Figure 2.2 and Figure 3.1 indicated that there is a wide range of requirements engineering and modeling techniques, there is far more to requirements engineering than a collection of techniques. In "Requirements Engineering Patterns" (May 2000, Section 3.5.3), Scott Ambler describes a collection of best practices, in the form of patterns (see Luke Hohmann's "Getting Started with Patterns" in Section 2.1.4), applicable to the Requirements workflow. His *Requirements First* pattern states that developers must form a firm foundation based on identified requirements from which to perform their work. Two of his other patterns will come in handy on any project beset by politics — *Requirements as Stick* and *Requirements as Shield*. The *Requirements as Stick* pattern describes how to use a well-defined set of requirements to evaluate alternative approaches or strategies to modify, or even stop, alternatives that your team finds undesirable. Going down the wrong technical path can seriously harm your project and this pattern provides a way to avoid doing so. The complement to this pattern is *Requirements as Shield*, which you apply to justify your approach to solving a given problem when it comes under scrutiny from other factions within your organization. By being able to show that your team is on the right technical path, you can garner the support of your project stakeholders — an important goal of the Inception phase.

There is far more to the Requirements workflow than simply creating a collection of models.

3.2 Requirements Gathering Techniques

Joint Application Development (JAD), also known as Joint Application Design, is a collaborative technique where developers and project stakeholders work together in facilitated meetings. JAD sessions can greatly improve your requirements gathering and overall product quality, but to make JAD work, you need to plan how you're going to use it in your organization. Jim Geier overviews the JAD technique in "Don't Get Mad, Get JAD!" (March 1996, Section 3.5.4). JAD sessions are basically structured meetings that are led by a trained facilitator and are often held to gather requirements, to design part of, or the whole, application, or to perform walkthroughs of a project deliverable. In a requirements JAD, the facilitator will often lead the group through techniques such as CRC modeling, essential modeling (Constantine & Lockwood, 1999; Ambler, 2001), and/or the definition of use cases. Geier argues that JAD is a technique that highly improves the process of developing software, especially the formation of requirements that lead to quality systems and products that are on time and within budget. However, as with everything else, JAD is not a silver bullet that will immediately cure all of your software development problems. JAD is a best practice that is part of the overall solution, not the entire solution.

Requirement JAD sessions are structured meetings where a facilitator leads developers and project stakeholders through the development of your requirements-oriented artifacts.

Every system has a collection of rules and constraints that it must support and conform to. These rules and constraints are requirements for your system, that should be documented in your supplementary specification (Kruchten, 2000). Business rules are the policies and constraints of the business, and are effectively what a functional requirement "knows" — the decisions, guidelines, and controls that are behind that functionality. In "Capturing Business Rules" (January 2000, Section 3.5.5), Ellen Gottesdiener shows that to support our users with good software, you need to understand the rules of logic behind the businesses that your software addresses. Gottesdiener provides the guidance that software developers need to be successful at the identification and documentation of business rules — a topic that is, unfortunately, glossed over in much of the literature regarding the Unified Modeling Language (UML) and the Unified Process. Until now.

Every system has a collection of pertinent business rules that should be documented in your supplementary specification.

3.3   User Interfaces and Internationalization

An important part of understanding the requirements of a system is to understand how users will work with your system. One way to accomplish this is to develop a prototype of the system with your users, often starting with an essential prototype (Constantine & Lockwood, 1999; Ambler, 2001) and evolving it over time, into a full-fledged user interface prototype. The implication is that requirements engineers must have at least a basic understanding of user interface development — an important aspect of which is usability. In "Learning the Laws of Usability" (October 1999, Section 3.5.6), Lucy Lockwood, co-author with Larry Constantine of *Software for Use* (1999), presents valuable guidelines — such as designing for real work, not getting concrete too quickly, and striving for efficient interaction — to help improve the quality of your prototypes. She believes that careful attention to the user interface is a hallmark of high-quality software systems and that a well-designed user interface makes the user's interaction with the software more efficient, less prone to error, faster to learn, and more productive.

Requirements engineers need at least a basic understanding of the fundamentals of user interface development.

Modern-day software, particularly that deployed to the Internet, is used by a wide range of users that are often working in different locales across the world. During the Inception phase, when you are identifying the initial requirements for your system, you must take the time to determine what your internationalization (i18n) requirements are and act accordingly. Taking i18n issues into account late in a project, during the Transition phase, or even during the Construction phase (Ambler & Constantine, 2000b), is often a recipe for disaster, as you will discover in Benson I. Margulies' "Your Passport to Proper Internationalization" (May 2000, Section 3.5.7). There are many potential procedural and technical pitfalls in building international software, and Margulies argues that there is no substitute for thinking carefully about the problem and formulating a plan of attack. He starts by describing a project that ran aground with their i18n efforts. This team made several classic mistakes. First, they made no attempt to verify that the engineering was being done to provide them a product to deliver to other countries. Second, they underestimated the scope of i18n by only expecting some minor visual blemishes and inaccuracies. Instead, they discovered that they suffered from display corruption, mysterious database errors, and poor language translation. To avoid internationalization problems, Margulies's experience is that you must consider your organizational structure, process and technical requirements and that you need to recognize the fact that there is no such thing as "localizing" a body of code that has never been internationalized before. Finally, he presents an overview of the techniques and technologies of i18n, showing that there is far more to it than simply using double-byte character sets (DBCS), such as Unicode, and hoping for the best.

To succeed at internationalization, you must consider your organizational structure, as well as process and technical requirements.

In "Thirteen Steps to a Successful System Demo" (March 1995, Section 3.5.8), Larry Runge describes a collection of best practices and insights for demonstrating software — a skill that is every bit as important as developing an effective user interface. His thirteen steps are:

1. Budget your time
2. Make a list of key points
3. Prioritize your points
4. Determine key features
5. Develop a script
6. Add detail
7. Make brief notes
8. Make a dry run
9. Expand or contract based on dry run
10. Prioritize the demonstration
11. End effectively
12. Dress professionally
13. Plan for the worst

3.4 Lessons from the Real World

We wrap up this chapter with Larry Constantine's "Real Life Requirements" (May 1998, Section 3.5.9), a collection of insights from everyday software developers. The article summarizes lessons from the real world that came out of the first *Management Forum — Live!* at the *Software Development '98 Conference and Exposition* held in San Francisco. The article addresses issues such as:

- understanding the difference between what users want and what users need (which is key to quality software development),
- coping with feature/scope creep,
- scoping requirements to fit within schedule constraints,
- fixing requirements too rigidly, too early, or too thoroughly(which can be as much of a problem as giving them short shrift),
- understanding your customers' definitions of success,
- defining and reviewing requirements with clients and users,
- educating your clients and users regarding costs and trade-offs,

- concentrating on the core requirements,
- building client comfort and confidence through reliable revisions and releases,
- distinguishing drivers from constraints, and
- distinguishing between what is critical and what is decorative.

Ultimately, Constantine shows that we are in the business of delivering solutions to real problems, and that requires us to give clients what they need more than what they want or claim to want. The most important qualities are those that let users accomplish work, making it easier, faster, and more valuable to the organization.

3.5   The Articles

3.5.1 "Decoding Business Needs" by Ellen Gottesdiener
3.5.2 "Customer Rights and Responsibilities" by Karl Wiegers
3.5.3 "Requirements Engineering Patterns" by Scott W. Ambler
3.5.4 "Don't Get Mad, Get JAD!" by Jim Geier
3.5.5 "Capturing Business Rules" by Ellen Gottesdiener, edited by Larry Constantine
3.5.6 "Learning the Laws of Usability" by Lucy Lockwood
3.5.7 "Your Passport to Proper Internationalization" by Benson I. Margulies
3.5.8 "Thirteen Steps to a Successful System Demo" by Larry Runge
3.5.9 "Real Life Requirements" by Larry Constantine

3.5.1 "Decoding Business Needs"

by Ellen Gottesdiener

Corporate software is designed to serve and support business needs, but how well it accomplishes its many missions often leaves much to be desired. Whether judged by internal measures (defects, process, or plan vs. delivery assessments) or external ones (surveys, metrics), it's clear that the industry falls far short of its ambitions.

In *Patterns of Software Systems Failure and Success* (Thomsen Computer Press, 1996), Capers Jones reports a high degree of risk in information technology projects, namely scope creep, schedule pressure and quality problems. The Standish Group's CHAOS report (www.standishgroup.com/chaos.html, 1995) shows that more than a quarter of software projects fail and that nearly half of all IT projects are "challenged" (completed, but over-budget, behind schedule, and delivering fewer features/functions than planned). Internal views of software production indicate high defect rates (see the Software Engineering Institute's Capability Maturity Model data at www.sei.cmu.edu/cmm/cmm.html) and immature software development processes.

The Standish Group suggests that the top four success factors for an IT project are, in ascending order: customer involvement, executive management support, clear statement of requirements and proper planning.

Active, visible business customer involvement is often elusive. Customers sponsor the project, specify requirements, test the product and use it after implementation. Yet developers are notoriously poor at establishing,maintaining and managing good customer relationships.

Facilitated workshops reduce the risk of scope creep
from 80% to 10%, accelerate the delivery of early lifecycle phases
by 30% to 40%, and providea 5% to 15% overall savings in time
and effort throughout the entire lifecycle.

Producing successful software depends on five important factors:

- Business staff must have formal roles in IT projects.
- Business staff must participate from the beginning.
- Developers should not proceed with a project without business sponsorship.
- If business involvement is a risk factor, it should be part of a risk-management strategy including strong communications and an organizational change management plan.
- Internal project processes and techniques must maximize customer collaboration.

There are three practical ways to involve business customers in the software development process: establishing a project charter, using facilitated workshops and addressing organizational change.

Table 3.2   The facilitation essentials method for managing workshops.

Purpose	Participants	Principles	Products	Place	Process
Why do we do things?	*Who* is involved?	*How* do we function?	*What* do we do?	*Where* is it located?	*When* do things happen?
• Goals	• People	• Guidelines	• Deliverables	• Venue	• Activities
• Needs	• Roles	• Ground rules	• Models	• Space	• Concurrency
• Motivation	• Players	• Process rules	• Decisions	• Time	• Sequence
• Justification	• Contributors		• Plans		• Order

The Commission

A charter begins with the end in mind, defining the who, what, when, where, why and how of the project. The charter documents:

- roles and responsibilities within the project organization (who),
- business functions in and out of scope, organizational scope, temporal scope, financial scope, and the priority of constraints such as time, cost, features/functions, quality, process and software metrics (what),
- the preliminary plan for the first phase of the project (when),
- where the work will get done and where the software will be deployed (where),
- goals and objectives (why), and
- risks and a risk-mitigation plan, methodologies, tools, assumptions, controls, a quality plan and knowledge transfer (how).

As Winston Churchill said, "The plan is nothing; the planning is everything." Indeed, the primary value of a project charter is not the resulting document but rather the process of creating, validating and closing the charter.

One of the roles that should be explicitly identified in the charter is the sponsor; that is, the person with the financial and logistical authority to make the project happen. The project manager must work with the sponsor to define what behaviors — actions and words — will be expected from the sponsor.

Depending on the organizational scope of the project, a more complex sponsorship team may be required. For example, I once participated in a global project that had an executive sponsor, a steering committee and a day-to-day project sponsor. In any case, the roles and responsibilities should be made clear to everyone connected with the project.

Displaying active, visible sponsorship entails making decisions, finding resources, promoting the project to peers and upper management and continually rewarding project members. Just as the act of specifying project risks tends to minimize them through team awareness, the very act of discussing, documenting and reviewing the expectations for the sponsor has the benefit of getting active sponsor participation. This simple activity is also a meaningful and important method of involving customers. Some sponsorship behaviors might include:

- defining and/or validating business rules and policies which are being implemented in the software,
- making decisions — and deciding how to make decisions — when multiple choices are possible,
- getting people resources to work on the project when they are needed,
- selling and marketing the value of the project,
- paying for the project or getting the money to pay for the project,
- making high-stake decisions quickly,
- ensuring that business objectives are being satisfied through the use of technology,
- ensuring that business objectives are being satisfied through the use of existing or new processes, and
- making the time to do all of the above.

In some projects, a written sponsorship contract can be useful as a concrete way for the team to ask for the behaviors it wants from the sponsor. If the sponsor is unable or unwilling to sign a sponsorship contract, it's better to know early. The project should not commence until sponsorship is secured.

How do you manage the chartering process? One way is to timebox the delivery of the charter in the kick-off phase initiated with a workshop for the project sponsor and the team. Prior to the workshop, portions of the charter should be drafted; decisions and definitions of missing or controversial pieces of the charter are made at the workshop.

In charter workshops that I have facilitated, the participants delivered:

- a list of functions that are in and out of scope,
- documented corrections or additions to the project roles and responsibilities,
- a visual scope model,
- a decision on the project sponsor,
- a list of risks,
- a ranking of risks and a risk-mitigation plan,
- a deliverables-based project plan,
- a statement of project strategy (project approach or method),

- a list of potential business analyses for the next phase,
- a project communications plan, and
- a decision on project constraints.

During the chartering stage, a sponsor must make decisions on roles/responsibilities, constraints and scope. But how does a team decide how it will decide?

Projects often get stuck when decisions are not made in a clear, timely manner. The sponsor is responsible for high-stakes decisions and must begin to exercise this authority early in the process. Using a decision rule and agreed-upon decision-making process is imperative.

At a recent business rules chartering workshop, I had to help the workshop sponsor define the deliverables of the encounter. One product was an intangible — a decision on the scope of business rules for the first phase of the project. As facilitator, I showed the sponsor a simple diagram of the possible decision rules she could exercise (leader decides after discussion, coin flip, delegation, majority rule, consensus, etc.). The sponsor is the leader, and thus is responsible for deciding how to decide a matter of project scope. (In other situations, the project manager or some other team role may inhabit this role.) Using a decision rule process, she could simultaneously check the participants' degree of agreement and model appropriate behavior.

To close out the chartering phase, conduct a walk-through workshop. This should result in a formal sign-off of the charter. If minor changes are still needed, the participants must agree on a sign-off date and responsibilities for those changes.

Facilitated Workshops

A facilitated workshop is a planned collaborative event in which participants, led by a neutral guide, deliver products in a concentrated period of time. Prior to the workshop, the participants agree upon what will be delivered and the ground rules for the interaction (including decision rules). The workshop process exploits the power of diverse groups of people joined together for a common goal. Participants act as a sophisticated team, working in a manner to achieve what psychologists call "consensual validation." A successful and productive workshop should be fun and energizing.

Facilitated workshops are most effective in the early stages of the software development lifecycle for chartering, planning, requirements, analysis and design. Not only can a well-run workshop provide the project artifacts (models, decisions, etc.) in a fast and high-quality manner, but it also has the benefit of building a team and establishing a spirit of real collaboration among all team members. Therefore, facilitated mid-point (or periodic) and debrief workshops are essential to team and project process improvement. Since business customers are participants in these workshops, and they provide about 19 percent of the total effort on IT projects, planning, requirements and debriefing workshops set the stage for and maintain active customer involvement.

Workshops for IT projects have their roots in Joint Application Design (JAD), a workshop technique developed and trademarked by IBM in the late 1970s. Since then, the process has evolved and been tailored to a variety of project types and technologies. The data around the increased quality and reduced costs of using JAD-like workshops are impressive.

Table 3.3   Products of the chartering/planning process.

- Business goals and objectives
- Problem/opportunity fishbone diagram
- Relationship map
- Context diagram, scope diagram
- Event list
- High-level business process map
- Prioritized constraints (cost, time, features/functions, quality)
- Project critical success factors/assumptions
- Customer types and needs (QFD)
- Future changes/barriers
- Action plan, pert or chart
- Gantt chart
- Risks/risk mitigation strategy
- Sponsor
- Project roles, responsibilities and organization
- Business policies

According to Capers Jones, while 60% of software defects originate in the requirements and design phases, early facilitated workshops reduce those defects by 20% to 60%. Facilitated workshops reduce the risk of scope creep from 80% to 10%, accelerate the delivery of early lifecycle phases by 30% to 40%, and provide a 5% to 15% overall savings in time and effort throughout the entire lifecycle, writes Jones in *Assessment and Control of Software Risk* (Prentice Hall, 1994). Workshops are powerful devices to deliver artifacts of project chartering, planning and requirements/analysis phases (see Table 3.3 and Table 3.4 for examples of some of the artifacts that can delivered in workshops.) After going through team-development stages such as "forming, storming, norming and performing," a group can become very productive, very fast. The following are real examples:

- In a business-rule workshop, participants delivered 35 business rules at the end of the third workshop day (averaging nearly 12 per day), then were able to add an additional 20 by the end of day four and an additional 35 business rules by the end of day five.
- In a use case modeling workshop, business participants were able to test their use case model and structural business rules using 55 scenarios in less than 90 minutes.
- In two hours, workshop participants generated 119 business events, classified them and then removed those not in scope.
- In three hours, a team validated a medium-complexity data model and made decisions about the scope of the data about which business performance metrics would be based.
- In 3.5 hours, a complete set of relationship maps were built by workshop participants which enabled them to identify, in detail, current business process and problem areas.
- In 75 minutes, a chartering workshop group delivered and categorized nine risk areas and created 13 risk-mitigation strategies for the two high-probability/high-impact risks.

How is this achieved? Planning is all. A facilitated workshop must be well-designed and planned in order to be successful, and it requires a process. The Facilitated Essentials method employs the Total Quality Management cycle of "plan, do, check, act" (PDCA), ensuring that the process is continually working. For example, a workshop contract, sometimes in the form of an agenda, will delineate decision rules and products. In other cases, an orientation is conducted to ensure agreement on the workshop process and understanding of the products and pre-work.

A framework for the Facilitation Essentials method (see Table 3.2), helps manage the quality of the process. Like John Zachman's framework for information systems architecture (*IBM Systems Journal*, vol. 26, no. 3, 1987), these columns represent all the interrogatives necessary to plan a successful workshop. Attention to all these dimensions is paramount. Customers are involved from the start of the workshop-planning process, beginning with identifying the workshop goals. This promotes a shared sense of purpose for the workshop and the project.

Table 3.4   Products of requirements analysis.

- Business process thread
- Process dependency diagram
- Business and temporal events
- Role/event matrix
- Event partitioned dataflow
- Activity value analysis
- CRCs, class model
- Cross-functional process map
- Scenarios
- Swim-lane/process map
- Use cases
- Window hierarchy diagram
- Object states
- Interface prototype
- Entity-relationship model
- Non-functional requirements list
- Entity life history diagram
- Requirements validation action plan
- Statechart diagram
- Business rule list
- Functional decomposition diagram

Organizational Change Management

Software modifies behavior: users navigate new work processes and procedures, performance expectations change — yet scant attention is paid to these post-implementation issues. Developers are in a unique position to improve the probability of success by paying attention to these aspects of change.

Organizational change management (OCM) is, in fact, what software development efforts are designed to do. The business goals and objectives which drive development are based on the expectation of change. So it's not surprising that when software projects fail or disappoint people, it often has nothing to do with the technology and everything to do with people factors — environment, rewards, feedback, procedures, measures, work aids, communications, etc. Here are some ways of mollifying customers undergoing a metamorphosis:

- Establish an OCM plan, or build it into the project plan. This includes a timeline for the change, metrics and who will act as sponsors (perhaps the project sponsor or other strategic individuals); change agents (for example, business analysts who participate in requirements workshops); and change targets (end users) who will participate in various life-cycle stages such as requirements, testing, prototyping, training and documentation.

- Identify the "end state" of the business environment during the chartering phase: How work will change, what procedures and methods in the business process will be altered, who will be affected, how the work flow might look, what metrics will be applied to the value of software, etc.

- Establish a communication plan for the project; include ongoing feedback mechanisms in this plan.

- Identify changes that can be expected during and after implementation, and identify the agents, targets, and sponsors (these should map to roles in the charter).

- Encourage your business partners to assess their readiness for change; this might mean stopping a project, revisiting project goals and objectives or taking other business actions to prepare for change.

- Establish a sponsorship contract for the change in which the specific behaviors which change sponsors must express (say), model (do) and reinforce (reward) are delineated; these contracts are established down the chain of influence in the organization to reach the ultimate targets of the change, e.g. the end users.

An Orchestrated Event

These three activities — creating a project charter, using facilitated workshops and addressing organization change — are highly interrelated. Software development is part of an orchestrated business change event. Delivering the wrong software correctly or delivering the right software incorrectly can make or break not only the project, but also the business. Satisfied customers are the product of continual business involvement in the life cycle.

3.5.2 "Customer Rights and Responsibilities"

by Karl Wiegers

Software success depends on developing a collaborative partnership between software developers and their customers. Too often, though, the customer-developer relationship becomes strained or even adversarial. Problems arise partly because people don't share a clear understanding of what requirements are and who the customers are. To clarify key aspects of the customer-developer partnership, I propose a Requirements Bill of Rights for software customers and a corresponding customer's Requirements Bill of Responsibilities. And, because it's impossible to identify every requirement early in a project, the commonly used — and sometimes abused — practice of requirements sign-off bears further examination.

Who is the Customer?

Anyone who derives direct or indirect benefit from a product is a customer. This includes people who request, pay for, select, specify or use a software product, as well as those who receive the product's outputs. Customers who initiate or fund a software project supply the high-level product concept and the project's business rationale. These business requirements describe the value that the users, developing organization or other stakeholders want to receive from the system.

The next level of requirements detail comes from the customers who will actually use the product. Users can describe both the tasks they need to perform — the use cases — with the product and the product's desired quality attributes. Analysts (individuals who interact with customers to gather and document requirements) derive specific software functional requirements from the user requirements.

Unfortunately, customers often feel they don't have time to participate in requirements elicitation. Sometimes customers expect developers to figure out what the users need without a lot of discussion and documentation. Development groups sometimes exclude customers from requirements activities, believing that they already know best, will save time or might lose control of the project by involving others. It's not enough to use customers just to answer questions or to provide selected feedback after something is developed. Your organization must accept that the days of shoving vague requirements and pizzas under the door to the programming department are over. Proactive customers will insist on being partners in the venture.

For commercial software development, the customer and user are often the same person. Customer surrogates, such as the marketing department, attempt to determine what the actual customers will find appealing. Even for commercial software, though, you should get actual users involved in the requirements-gathering process, perhaps through focus groups or by building on your existing beta testing relationships.

The Customer-Development Partnership

Quality software is the product of a well-executed design based on accurate requirements, which are in turn the result of effective communication and collaboration — a partnership — between developers and customers. Collaborative efforts only work when all parties involved know what they need to be successful, and when they understand and respect what their collaborators need to succeed. As project pressures rise, it's easy to forget that everyone shares a common objective: To build a successful product that provides business value, user satisfaction and developer fulfillment.

A shared understanding of the requirements approval process should alleviate the friction that can arise as the project progresses and oversights are revealed, or as marketplace and business demands evolve.

Table 3.5 A Bill of Rights for software customers.

As a software customer, you have the right to:
1. Expect analysts to speak your language.
2. Expect analysts to learn about your business and your objectives for the system.
3. Expect analysts to structure the requirements information you present into a software requirements specification.
4. Have developers explain requirements work products.
5. Expect developers to treat you with respect and to maintain a professional attitude.
6. Have analysts present ideas and alternatives both for your requirements and for implementation.
7. Describe characteristics that will make the product easy and enjoyable to use.
8. Be presented with opportunities to adjust your requirements to permit reuse.
9. Be given good-faith estimates of the costs and trade-offs when you request a change.
10. Receive a system that meets your functional and quality needs, to the extent that those needs have been communicated to the developers and agreed upon.

Table 3.5 presents a Requirements Bill of Rights for Software Customers, 10 expectations that customers can place on their interactions with analysts and developers during requirements engineering. Each of these rights implies a corresponding software developer's responsibility. Table 3.6 proposes 10 responsibilities the customer has to the developer during the requirements process. These rights and responsibilities apply to actual user representatives for internal corporate software development. For mass-market product development, they apply more to customer surrogates, such as the marketing department.

Table 3.6 Bill of Responsibilities for software customers.

As a software customer, you have the responsibility to:
1. Educate analysts about your business and define jargon.
2. Spend the time to provide requirements, clarify them, and iteratively flesh them out.
3. Be specific and precise about the system's requirements.
4. Make timely decisions about requirements when requested to do so.
5. Respect developers' assessments of cost and feasibility.
6. Set priorities for individual requirements, system features, or use cases.

7. Review requirements documents and prototypes.

8. Promptly communicate changes to the product's requirements.

9. Follow the development organization's defined requirements change process.

10. Respect the requirements engineering processes the developers use.

Early in the project, customer and development representatives should review these two lists and reach a meeting of the minds. If you encounter some sticking points, negotiate to reach a clear understanding regarding your responsibilities to each other. This understanding can reduce friction later, when one party expects something the other isn't willing or able to provide. These lists aren't all-inclusive, so feel free to change them to meet your specific needs.

Customer Rights

Right #1: To expect analysts to speak your language. Requirements discussions should center on your business needs and tasks, using your business vocabulary (which you might have to convey to the analysts). You shouldn't have to wade through computer jargon.

Right #2: To expect analysts to learn about your business. By interacting with users while eliciting requirements, the analysts can better understand your business tasks and how the product fits into your world. This will help developers design software that truly meets your needs. Consider inviting developers or analysts to observe what you do on the job. If the new system is replacing an existing application, the developers should use the system as you do to see how it works, how it fits into your workflow, and where it can be improved.

Right #3: To expect analysts to write a software requirements specification (SRS). The analyst will sort through the customer-provided information, separating actual user needs from other items, such as business requirements and rules, functional requirements, quality goals and solution ideas. The analyst will then write a structured SRS, which constitutes an agreement between developers and customers about the proposed product. Review these specifications to make sure they accurately and completely represent your requirements.

Right #4: To have developers explain requirements work products. The analyst might represent the requirements using various diagrams that complement the written SRS. These graphical views of the requirements express certain aspects of system behavior more clearly than words can. Although unfamiliar, the diagrams aren't difficult to understand. Analysts should explain the purpose of each diagram, describe the notations used, and demonstrate how to examine it for errors.

Right #5: To expect developers to treat you with respect. Requirements discussions can be frustrating if users and developers don't understand each other. Working together can open each group's eyes to the problems the other faces. Customers who participate in requirements development have the right to have developers treat them with respect and to appreciate the time they are investing in project success. Similarly, demonstrate respect for the developers as they work with you toward your common objective of a successful project.

Right #6: To have analysts present ideas and alternatives for requirements and implementation. Analysts should explore ways your existing systems don't fit well with your current business processes, to make sure the new product doesn't automate ineffective or inefficient processes. Analysts who thoroughly understand the application domain can sometimes suggest improvements in your business processes. An experienced and creative analyst also adds value by proposing valuable capabilities the new software could provide that the users haven't even envisioned.

Right #7: To describe characteristics that will make the product easy and enjoyable to use. The analyst should ask you about characteristics of the software that go beyond your functional needs. These "quality attributes" make the software easier or more pleasant to use, letting you accomplish your tasks accurately and efficiently. For example, customers sometimes state that the product must be "user-friendly" or "robust" or "efficient," but these terms are both subjective and vague. The analyst should explore and document the specific characteristics that signify "user-friendly," "robust," or "efficient" to the users.

Right #8: To be presented with opportunities to adjust your requirements to permit reuse. The analyst might know of existing software components that come close to addressing some need you described. In such a case, the analyst should give you a chance to modify your requirements to allow the developers to reuse existing software. Adjusting your requirements when sensible reuse opportunities are available can save time that would otherwise be needed to build precisely what the original requirements specified.

Right #9: To be given good-faith estimates of the costs of changes. People sometimes make different choices when they know one alternative is more expensive than another. Estimates of the impact and cost of a proposed requirement change are necessary to make good business decisions about which requested changes to approve. Developers should present their best estimates of impact, cost, and trade-offs, which won't always be what you want to hear. Developers must not inflate the estimated cost of a proposed change just because they don't want to implement it.

Right #10: To receive a system that meets your functional and quality needs. This desired project outcome is achievable only if you clearly communicate all the information that will let developers build the product that satisfies your needs, and if developers communicate options and constraints. State any assumptions or implicit expectations you might hold; otherwise, the developers probably can't address them to your satisfaction.

Customer Responsibilities

Responsibility #1: To educate analysts about your business. Analysts depend on you to educate them about your business concepts and terminology. The intent is not to transform analysts into domain experts, but to help them understand your problems and objectives. Don't expect analysts to have knowledge you and your peers take for granted.

Responsibility #2: To spend the time to provide and clarify requirements. You have a responsibility to invest time in workshops, interviews and other requirements elicitation activities. Sometimes the analyst might think she understands a point you made, only to realize later that she needs further clarification. Please be patient with this iterative approach to developing and refining the requirements, as it is the nature of complex human communication and essential to software success.

Responsibility #3: To be specific and precise about requirements. Writing clear, precise requirements is hard. It's tempting to leave the requirements vague, because pinning down details is tedious and time-consuming. At some point during development, though, someone must resolve the ambiguities and imprecisions. You are most likely the best person to make those decisions; otherwise, you're relying on the developers to guess correctly. Do your best to clarify the intent of each requirement, so the analyst can express it accurately in the SRS. If you can't be precise, agree to a process to generate the necessary precision, perhaps through some type of prototyping.

Responsibility #4: To make timely decisions. The analyst will ask you to make many choices and decisions. These decisions include resolving inconsistent requests received from multiple users and making trade-offs between conflicting quality attributes. Customers who are authorized to make such decisions must do so promptly when asked. The developers often can't proceed until you render your decision, so time spent waiting for an answer can delay progress. If customer decisions aren't forthcoming, the developers might make the decisions for you and charge ahead, which often won't lead to the outcome you prefer.

Responsibility #5: To respect a developer's assessment of cost and feasibility. All software functions have a price, and developers are in the best position to estimate those costs. Some features you would like might not be technically feasible or might be surprisingly expensive to implement. The developer can be the bearer of bad news about feasibility or cost, and you should respect that judgment. Sometimes you can rewrite requirements in a way that makes them feasible or cheaper. For example, asking for an action to take place "instantaneously" isn't feasible, but a more specific timing requirement ("within 50 milliseconds") might be achievable.

Responsibility #6: To set requirement priorities. Most projects don't have the time or resources to implement every desirable bit of functionality, so you must determine which features are essential, which are important to incorporate eventually, and which would just be nice extras. Developers usually can't determine priorities from your perspective, but they should estimate the cost and technical risk of each feature, use case, or requirement to help you make the decision.

When you prioritize, you help the developers deliver the greatest value at the lowest cost. No one likes to hear that something he or she wants can't be completed within the project bounds, but that's just a reality. A business decision must then be made to reduce project scope based on priorities or to extend the schedule, provide additional resources or compromise on quality.

Responsibility #7: To review requirements documents and prototypes. Having customers participate in formal and informal reviews is a valuable quality control activity — indeed, it's the only way to evaluate whether the requirements are complete, correct and necessary.

It's difficult to envision how the software will actually work by reading a specification. To better understand your needs and explore the best ways to satisfy them, developers often build prototypes. Your feedback on these preliminary, partial or possible implementations helps ensure that everyone understands the requirements. Recognize, however, that a prototype is not a final product; allow developers to build fully functioning systems based on the prototype.

Responsibility #8: To promptly communicate changes to the product's requirements. Continually changing requirements pose a serious risk to the development team's ability to deliver a high-quality product within the planned schedule. Change is inevitable, but the later in the development cycle a change is introduced, the greater its impact.

Extensive requirements changes often indicate that the original requirements elicitation process wasn't adequate.

Changes can cause expensive rework and schedules can slip if new functionality is demanded after construction is well under way. Notify the analyst with whom you're working as soon as you become aware of any change needed in the requirements. Key customers also should participate in the process of deciding whether to approve or reject change requests.

Responsibility #9: To follow the development organization's requirements change process. To minimize the negative impact of change, all participants must follow the project's change-control process. This ensures that requested changes are not lost, the impact of each requested change is evaluated, and all proposed changes are considered in a consistent way. As a result, you can make good business decisions to incorporate certain changes into the product.

Responsibility #10: To respect the requirements engineering processes the developers use. Gathering requirements and verifying their accuracy are among the greatest challenges in software development. Although you might become frustrated with the process, it's an excellent investment that will be less painful if you understand and respect the techniques analysts use for gathering, documenting, and assuring the quality of the software requirements. Customers should be educated about the requirements process, ideally attending classes together with developers. I've often presented seminars to audiences that included developers, users, managers, and requirements specialists. People can collaborate more effectively when they learn together.

What About Sign-Off?

Agreeing on a new product's requirements is a critical part of the customer-developer partnership. Many organizations use the act of signing off on the requirements document to indicate customer approval. All participants in the requirements approval process need to know exactly what sign-off means.

One potential problem is the customer representative who regards signing off on the requirements as a meaningless ritual: "I was given a piece of paper that had my name printed beneath a line, so I signed on the line because otherwise the developers wouldn't start coding." This attitude can lead to future conflicts when that customer wants to change the requirements or when he's surprised by what is delivered: "Sure, I signed off on the requirements, but I didn't have time to read them all. I trusted you guys — you let me down!"

Equally problematic is the development manager who views sign-off as a way to freeze the requirements. Whenever a change request is presented, he can point to the SRS and protest, "You signed off on these requirements, so that's what we're building. If you wanted something else, you should have said so."

Both of these attitudes fail to acknowledge the reality that it's impossible to know all the requirements early in the project and that requirements will undoubtedly change over time. Requirements sign-off is an appropriate action that brings closure to the requirements development process. However, the participants have to agree on precisely what they're saying with their signatures.

More important than the sign-off ritual is the concept of establishing a "baseline" of the requirements agreement — a snapshot at some point in time. The subtext of a signature on an

SRS sign-off page should therefore read something like: "I agree that this document represents our best understanding of the requirements for the project today. Future changes to this baseline can be made through the project's defined change process. I realize that approved changes might require us to renegotiate the project's costs, resources and schedule commitments."

A shared understanding of the requirements approval process should alleviate the friction that can arise as the project progresses and requirements oversights are revealed, or as marketplace and business demands evolve. Sealing the initial requirements development activities with such an explicit agreement helps you forge a continuing customer-developer partnership on the way to project success.

3.5.3 "Requirements Engineering Patterns"

by Scott W. Ambler

Three approaches to motivate your developers to invest time
in taking care of first things first.

Requirements engineering — the elicitation, documentation and validation of requirements — is a fundamental aspect of software development. Unfortunately, and to the detriment of everyone involved, requirements engineering efforts are often shortchanged or even completely forgone in favor of "getting the code out as soon as possible, at all cost." This month I share my experiences and observations regarding requirements engineering, present several requirements engineering patterns that you should find valuable to your software development efforts, and suggest several good books that describe how to successfully engineer requirements.

First, let's define several important terms. A pattern is a solution to a common problem, taking relevant forces into account and enabling the reuse of proven techniques and strategies. Patterns can be described in template form (three patterns are presented in template form in the sidebar) or in degenerate form (that is, in prose). A pattern language is a collection of patterns that together describe how to completely address a large problem space, such as requirements engineering. A software process is the definition of the steps to be taken, the roles of the people performing those steps, and the artifacts being created or used by an organization to develop, operate and support a software product.

How do you go about defining requirements? In "Requirements by Pattern" (Dec. 1999), Christopher Creel, of TRC Inc., presented three patterns — Specify, Presentation and Prioritize — that encapsulate three strategies for documenting requirements. The intent of the Specify requirement pattern is to specify how an actor can identify an object, reducing the coupling between the method to find an object and the operations that are performed on that object. For example, in Windows Explorer, selecting its name from a list specifies a file. The intent of the Presentation requirement pattern is to describe the data that an application must present, enabling you to focus on the information to display, not how to display it.

Creating a Summary Report

Thus, to create a customer summary report, you would indicate that it presents the customer's name, phone number and address; you would not state that the customer name is shown in an Acme Corp. text widget for Java. The intent of the Prioritize requirement pattern is to communicate the urgency of one application aspect over another without specifying the exact implementation for communicating that priority. For example, items in a list may be color-coded to indicate priority or sorted in a defined order. Although these patterns do not form the basis for a complete requirements engineering pattern language, they do reveal that patterns are a viable approach for describing common approaches to requirements engineering.

Why requirements engineering? My experience is similar to Suzanne and James Robertson's statement in *Mastering the Requirements Process* (ACM Press, 1999): "If you do not have the correct requirements, you cannot design or build the correct product, and consequently the product does not enable the users to do their work." In fact, this concept is one of the main motivational factors behind the Requirements First pattern described in the sidebar. Requirements engineering is a fundamental aspect of traditional software processes — such as the waterfall approach — and of modern-day processes, such as the Unified Process. Figure 3.2 depicts the enhanced life cycle for the Unified Process, described in greater detail in *The Unified Process Elaboration Phase* (CMP Books, 2000), and, as you see, its Requirements workflow is an important aspect of the Inception, Elaboration, Construction and Transition phases of a software project. The implication is that you ignore requirements engineering at your own peril.

Figure 3.2 The enhanced life cycle for the Unified Process.

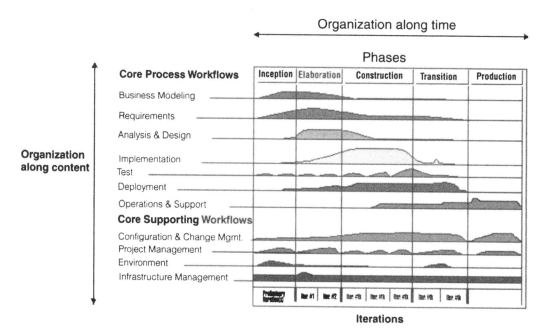

There is more to requirements engineering than its technical aspects — the politics are just as crucial to your success. I would love to live in a world in which I could focus on the technical nuances of a software development project, but, unfortunately, I just can't seem to find that mystical nirvana that all techies seem to seek.

Competing for Resources

Every project on which I have worked has had to compete for resources, fend off "attacks" from groups with different agendas and, more often than not, work through its own internal team politics. Over time, I have applied two patterns to overcome a wide range of political challenges: Requirements As Shield and Requirements As Stick. I apply Requirements As Shield whenever someone is criticizing the work of my team, ideally showing that the criticisms are not relevant to the problem at hand, based on our defined requirements. Sometimes we discover that our requirements are not complete, which is likely what my team's detractors were hoping for. However, this is a good result, because we want to discover any defects early in our project so we can correct them while they are still inexpensive to fix.

Refer to Steve McConnell's *Rapid Development* (Microsoft Press, 1996) or my own *Process Patterns* (Cambridge University Press, 1998) for discussions of the rising cost of fixing defects the later they are found.

Conversely, I apply Requirements As Stick to counteract any approaches or strategies proposed by developers or teams that I believe are not in the best interest of my organization: By evaluating these proposals against accepted requirements, you can quickly see any weaknesses.

I apply this pattern so often that my coworkers know they are in trouble whenever I ask the question, "So, do you have any requirements for that?" Taken with Creel's initial requirements patterns, we have a bare-bones start at a pattern language for requirements engineering. Our hope is that others will add to this collection of patterns over time, evolving it into a full-fledged pattern language. In fact, feel free to e-mail me at scott@ambysoft.com and Chris at Chris.Creel@trcinc.com with any suggestions.

Good References Available

Fortunately, there are many good requirements engineering books available. The book I would suggest is *Software For Use* by Larry Constantine and Lucy Lockwood (ACM Press, 1999), which focuses on usage-centered design techniques such as essential user-interface prototyping and essential use case modeling.

A close second is *Software Requirements* by Karl Wiegers (Microsoft Press, 1999), in which he presents a broad overview of requirements engineering best practices.

As I mentioned, I also like Suzanne and James Robertson's *Mastering The Requirements Process and Managing Software Requirements* by Dean Leffingwell and Don Widrig (Addison Wesley Longman, 2000); both books are good resources for anyone interested in learning about a wide range of new or traditional requirements techniques.

Finally, to learn more about use case modeling in particular, I highly suggest *Applying Use Cases* by Geri Schneider and Jason P. Winters (Addison Wesley Longman, 1998) and *Use Case Driven Object Modeling With UML* by Doug Rosenberg and Kendall Scott (Addison Wesley Longman, 1999). Both books are short, to the point and easy to understand.

Requirements engineering is an important and fundamental aspect of software development. Regardless of the techniques you employ — interviewing, use case modeling, essential prototyping, Class Responsibility Collaborator (CRC) modeling — the basics still remain the same.

You should start any development or development-related effort by identifying requirements: those that you will not only use to ensure that you are building or buying the right thing, but also those that will help you overcome the political roadblocks that you'll inevitably encounter.

AUTHOR'S NOTE: Special thanks to Chris Creel, who provided several key insights for this column.

Patterns for Requirements Engineering

Over the past few years, I have found three patterns — Requirements First, Requirements as Shield and Requirements as Stick — to be of great value in motivating fellow developers to invest their time in engineering requirements. An excellent discussion of pattern templates can be found in *The Patterns Handbook*, edited by Linda Rising (Cambridge University Press, 1998), and through links posted at The Process Patterns Resource Page at http://www.ambysoft.com/processPatternsPage.html.

Pattern: Requirements First

Intent. The intent of Requirements First is for developers to form a firm basis from which to perform their work. Before attempting to solve the problem at hand, developers must first identify the nature of the problem that they are trying to solve. In other words, start with requirements.

Initial Context. There is a nontrivial problem to be solved when using software or software-related artifacts. Examples include:

1. You must develop or purchase a new application to meet the needs of your user community.
2. You must develop or purchase a technical artifact, such as a framework or application server, to support your organization's development efforts.
3. You must purchase a development tool, such as a configuration management system or a modeling tool and make sure that it meets your needs.
4. You must develop a common corporate system architecture to support your organization's various software endeavors.

Motivation. You must understand what the software development problem is before you can solve it. A requirement describes one potential aspect of the problem to be solved, so you must first understand all the pertinent requirements that you are trying to fulfill.

Solution. Invest the time to define the problem to be solved, to engineer the requirements that your solution must fulfill and to engineer Requirements First.

Resulting context. Once the decision has been made to engineer requirements, you must do the actual work. Identify the sources of expertise, either people or information resources, as well as who can potentially verify the accuracy of the identified requirements. You may need someone to approve your requirements. Finally, you may need to train people in requirements engineering.

Pattern: Requirements As Stick

Intent. The intent of Requirements As Stick is to use a well-defined set of requirements to evaluate alternative approaches or strategies to modify or even stop undesirable alternatives.

Initial Context. You have identified and documented the pertinent requirements. There are always several ways to address any given problem, and each will have its supporters. Furthermore, there are always competing factions that support strategies that conflict with your goals. Finally, you must justify why these other approaches will not work. Examples include:

1. You must compare and contrast two approaches.

2. Another group within your organization is promoting an architecture that you believe will not work well with your current approach and that does not fully meet the needs of your organization.

3. An external consulting company proposes to develop a system for your organization for less than it could be developed for internally. You fear that the consultants have not fully considered the problem at hand or are purposely misleading your organization to win the contract.

Motivation. You want your organization to adopt the best solution to address a problem and reduce the internal politics that hamper your efforts to move forward.

Solution. Use your defined requirements as a stick to beat down any approach that is not viable, evaluating it against your requirements to understand its weaknesses.

Resulting context. You limit your range of choices to approaches that will actually solve the problem at hand. You also enable your organization to reduce its "political thrash" by focusing solely on the viable alternatives.

Pattern: Requirements As Shield

Intent. The intent of Requirements As Shield is to justify your defined approach to solving a given problem when it comes under scrutiny.

Initial Context. You have identified and documented the pertinent requirements. There are always several ways to address any given problem, and you have chosen one that you believe will work. Other factions, however, are criticizing your approach. You must justify why your strategy will work. Examples include:

1. Your design approach is criticized by another developer who is judging it against his own criteria.

2. You suggest using a new technology for the project that you are working on, and you must justify its adoption.

3. Your existing project is several months behind schedule and at risk of being assigned to another team, and you must justify why your approach is the best.

Motivation. You want your organization to adopt the best solution to address a problem and reduce the internal politics that hamper your efforts to move forward.

Solution. Use your requirements as a shield to protect against unwarranted criticisms. Counter each criticism by mapping it against your requirements, then from your requirements to your solution. Criticisms that are not applicable should be identified as such.

Resulting context. You will have either succeeded at justifying your approach or the criticisms will have revealed missing or misunderstood requirements, which are problems that you will need to address (assuming your effort hasn't been canceled).

— Scott Ambler

3.5.4 "Don't Get Mad, Get JAD!"

by Jim Geier

Joint Application Design can greatly improve your requirements gathering and overall product quality. But to make JAD work, you need to plan how you're going to use it in your organization.

Requirements are crucial to all development projects. They provide the basis for design, implementation, and support of the system or product. Many studies and testimonials prove that poor requirements are the leading cause of unsuccessful software projects. Requirements that are ambiguous, untestable, and most of all, not able to fully satisfy needs of potential users contribute to high development costs, lagging time-to-market, and unhappy users and customers.

Poor requirements are the basis for 60% to 80% of system defects that eventually surface late in the development phase or after delivery to users. These defects are very time-consuming and expensive to correct. Shabby requirements up front also lead to the continual stream of "new" requirements that often spring up throughout a project to make up for inadequacies in the system. This causes a great deal of rework, extending development time and costs. Organizations must emphasize the definition of requirements to keep their heads above water.

The Requirements Fumble

Many developers unknowingly do a sloppy job of determining requirements. This is often because of the development process their organizations follow, which is usually similar to the process shown in Figure 3.3. The customer represents the needs of potential users of the system or product under development. In companies that develop software products, the sales and marketing staff typically express customer needs in terms of requests and requirements. Otherwise, requirements generally flow directly from the customer. Project managers are often responsible for managing the overall development, installation, and support of the product or system. Typically they produce the first specification the development group uses to design and code the system or product.

A few problems are inherent in this process. First, the hand-offs between the different players can take a long time, delaying the creation of a prototype for validation purposes. Second, the customer isn't engaged in the entire process, which forces developers to guess what missing or incomplete requirements might be. Third, the true meaning of the requirements becomes lost as the project travels throughout the group. A simple game of telephone illustrates this point well.

Figure 3.3 Traditional communication flow during product development.

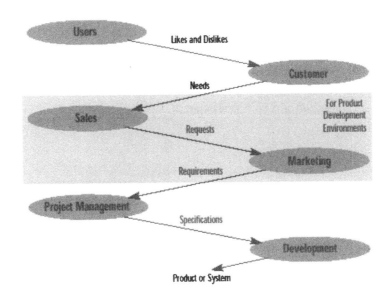

The process shown in Figure 3.3 can work well if the organization uses effective documentation to communicate ideas throughout the process. But how many organizations can truly admit they have documentation that clearly defines requirements? A different process is necessary.

JAD: The Bronze Bullet

JAD, Joint Application Design, is a technique that highly improves the process of developing software, especially the formation of requirements that lead to quality systems and products that are on time and within budget. JAD is not a silver bullet that will immediately cure your problems. The proper use of JAD requires careful planning and preparation to make it successful. Over time, though, most organizations can incorporate JAD and increase the success of their development projects.

JAD is a parallel process, simultaneously defining requirements in the eyes of the customer, users, sales people, marketing staff, project managers, and developers. The goal of JAD is for all team members — especially the customer and developers — to reach consensus on requirements, as shown in Figure 3.4. This is why JAD really pays off, ensuring the early definition of accurate requirements that will minimize later rework. Most projects can define requirements through JAD to provide enough basis for a prototype in days or weeks, rather than months.

Figure 3.4 The JAD approach for developing requirements.

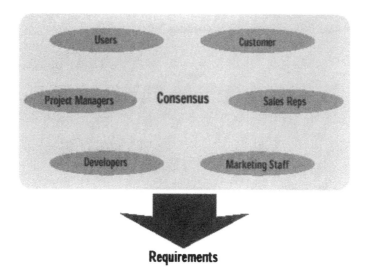

JAD is actually a meeting, or series of meetings, that have the customer and developers as active participants working together in the creation of requirements. The customer represents the potential users' functional areas, identifies business requirements, and makes final decisions on the content of requirements. Whenever feasible, potential users may also participate in the JAD to express their likes and dislikes. The developers are present to analyze system requirements based on customer needs.

In addition to the active participants, JAD includes a facilitator, a scribe, and optional observers. The facilitator manages the meeting and acts as a mediator or guide to guarantee the group stays focused on objectives and follows all the rules. The facilitator should have good communication skills, be impartial to the project, have team building experience and leadership skills, be flexible, and be an active listener.

The scribe records the proceedings of the JAD and should have good recording skills and some knowledge of the subject matter. It may be beneficial to have impartial observers to monitor the JAD sessions and provide feedback to the facilitator and project manager. In addition, managers as observers can spot and take action to problems that go beyond the scope of the facilitator's and project manager's domain. However, to ensure appropriate interaction among the customer and developers, observers must not actively participate during the JAD meeting.

Why does JAD improve requirements? Largely because the customer becomes a partner in the development project, allowing an effective customer-developer team. This teaming breaks down communication barriers, encourages communication, and increases levels of trust and confidence. Because JAD leads to requirements quickly, developers can start prototyping earlier. This is important because it provides a vision of the system for the users, fueling the refinement of requirements. JAD also keeps the customer accurately informed of what can and can't be done. Developers can validate the requirements as the customer states them — so marketing people can't promise things that developers can't develop within given constraints.

Implementing JAD

Here are a few tips for implementing JAD successfully:

1. *Make sure you have clear and appropriate objectives for the JAD meetings*. If not, the JAD proceedings will flounder and lead to unnecessary outcomes. For example, an objective for a JAD could be to define user interface requirements for a particular system.

2. *Obtain the appropriate level of coordination and commitment to using JAD*. The JAD facilitator in my organization, Peter Jones, who is writing a book on JAD, explains it this way: "In many cases, participation in a JAD will stretch across organizational boundaries. Developers are often from the information systems group, and the customer may represent users spanning several functional groups. Without concurrence of all group managers, the JAD meetings will appear biased to those not buying into the idea, causing some people to not participate or accept the outcomes. Therefore, to receive commitment to the method, the initiators of the JAD should discuss the benefits and purpose of the JAD with applicable managers of each group."

3. *Consider using an independent consultant as a facilitator*. This assures neutrality and avoids siding with one particular group of participants. Be certain, though, to avoid selecting a consultant closely allied with the department responsible for development. A close alliance here could tempt the facilitator to favor the developers, letting them dominate the meeting and hamper ideas from the customer. It doesn't hurt to have internal people in mind as facilitators, just be sure they have proper training and are not connected to the project they're facilitating.

4. *Talk to all participants before the JAD*. Discuss the issues involved with the particular project, such as potential conflicts. Give all new participants an orientation to JAD if it's their first time attending one. In some cases, it may be the first time business people and developers work together. Therefore, minimize communication problems by preparing participants to speak the same language. Avoid using computer jargon. Otherwise, communication may be difficult and the customer participation will decline.

5. *Establish rules*. This is absolutely necessary because the different agendas of the customer, users, and developers can often derail the JAD and raise conflicts. Rules should state that all members will conform to an agenda, all participants are equal, observers will remain silent, and that the bottom line is to reach consensus on requirements. Be sure to have the participants contribute to the formation of rules.

6. *Don't let developers dominate the meeting*. Many JADs tend to have too many developers and not enough representation of the potential users. This usually blocks users from expressing their ideas. Also, information systems departments tend to use JAD to "rubber stamp" the requirements; that is, to have the customer merely review and approve them. You should limit the developers to architects and engineers because programmers may push the team toward a design too soon. The facilitator must ensure that all participants have equal time to voice their ideas.

7. *Use effective communication tools*. As Peter Jones mentions, "Be certain to utilize a method that expresses ideas in a way that is understandable by both the customer and developers. The group should utilize a diagramming technique, such as data flow diagrams or simple flowcharts, to express requirements. This will make it easier for everyone to understand the requirements and make it possible to record them in a form the scribe can enter directly into a CASE tool."

8. *The facilitator shouldn't offer opinions, especially unsolicited ones.* If the facilitator acts as a consultant, it may squelch participation from the customer and users. Remember, the facilitator is there to keep the customer and users active.

3.5.5 "Capturing Business Rules"

by Ellen Gottesdiener, edited by Larry Constantine

An inept, inadequate or inefficient requirements analysis starts a project on the wrong foot. Time is wasted at a point in development when time is of the essence, and developers will find it hard to produce a good software product based on ill-defined requirements. Capturing, validating and verifying functional requirements are major challenges, not only for clients and business partners, but also for managers and software developers. We need clear and usable requirements that can guide the development of a quality software product, and a poor requirements process will not lead to a good requirements product.

The requirements process includes what techniques are used to capture the requirements (interviews, facilitated workshops, prototyping, focus groups or the like), what tools are used for capturing and tracing information (text, diagrams, narratives or formal models), and how customers and users are actively involved throughout the process. When it comes to the requirements product, management concerns include the testability of functional requirements, the ability to find and resolve conflicts in requirements, the ability to link requirements back to business goals (backwards traceability), and the ability to track functional requirements throughout the life cycle from design to code to test and deployment (forward traceability).

As mature managers, we no longer expect silver bullets for requirements engineering. However, in my experience there is at least one "secret weapon." Whatever functional models you wish to use, whether use cases, CRC cards or some proprietary technique championed by your boss, the important thing is to focus on the true essence of the functional requirements: the business rules.

Rules Rule

Business rules are the policies and constraints of the business, whether the "business" is banking, software development, or automation engineering. The Object Management Group simply defines them as "declarations of policy or conditions that must be satisfied." They are usually expressed in ordinary, natural language and are "owned" by the business. Your business may be commercial, not-for-profit, or part of a government, but it still has business rules. They provide the knowledge behind any and every business structure or process. Business rules are also, therefore, at the core of functional requirements. You may use various functional requirements models — structural (data, class); control-oriented (events, state-charts); object-oriented (classes/objects, object interactions); or process-oriented (functional decomposition, process threads, use cases, or the like), but within the heart of all functional requirements are business rules.

Business rules are what a functional requirement "knows" — the decisions, guidelines and controls that are behind that functionality. For example, functional requirements or models that include processes such as "determine product" or "offer discount" imply performance of actions, but they also embody knowledge — the underlying business rules —

needed to perform those actions. An insurance claim in a "pending" stage means that certain things (variables, links to other things) must be true (invariants, or business rules). The behavior of a claim in a pending lifecycle stage is, thus, dependent upon business rules which govern and guide behaviors.

As the essential ingredient of functional requirements, business rules deserve direct, explicit attention. Since business rules lurk behind functional requirements, they are easily missed and may not be captured explicitly. Without explicit guidance, software developers will simply make whatever assumptions are needed to write the code, thus building their assumptions about conditions, policies, and constraints into the software with little regard for business consequences. Rules that are not explicit and are not encoded in software through the guesses of developers may not be discovered as missing or wrong until later phases. This results in defects in those later phases that could have been avoided if the rules had been elicited, validated and baselined during requirements analysis. In the end, the lack of explicit focus on capturing the business rules creates rework and other inefficiencies.

Rather than just mentioning business rules as "notes" in your models, you should capture and trace them as requirements in and of themselves. Focusing on business rules as the core functional requirements not only promotes validation and verification, but it can speed the requirements analysis process.

Top-Down, Again

In working on a business-rule approach to software development, I have come to realize that such an approach needs to be driven from the top down, like the traditional top-down methods of information engineering. Unfortunately, in the modern world, with business moving at the speed of a mouse click, a systematic, top-down analysis is often an unaffordable luxury. Consequently, I've become more practical; a business-rules approach doesn't need to be a method or methodology in and of itself. Rather it is just a necessary part of requirements engineering. It includes a process (with phases, stages, tasks, roles, techniques, etc.), a business rule meta-model, and business rule templates.

Advocates of the recently standardized Unified Modeling Language (UML) and the accompanying all-purpose "Unified Process" toss around the term "business rules" in presentations and conversations, but neither UML nor UP offers much guidance for business rules. UML has an elaborate meta-model and meta-meta-model to support its language. One of the classes at the meta-model level is called "Rules." But the UML has given business rules short shrift. The only modeling element that is practically usable is the "constraint" element, which can be attached to any modeling element. Furthermore, although the UML's OCL (Object Constraint Language) is a fine specification-level language to attach business rules to structural models, it is not a language to use during requirements analysis when working directly with business customers to elicit and validate business rules. Business rules need to be treated as first-class citizens (no pun intended).

Use Cases and Business Rules

At the center of the UML are use cases, which are viewed by some analysts as business requirement models, while others call them "user-requirement models." Use cases are a functional (process) model that can be expressed as words using a natural language; some templates also may include special sections, such as pre- and post-conditions, goal name and the like. Use cases have proven to be an important and useful model for capturing requirements.

Recent work has evolved use cases from an often vague requirement deliverable into something specific, focused and usable. Alistar Cockburn has directed attention to the goals of use cases, and Larry Constantine and Lucy Lockwood have developed essential use cases for usage-centered design. However, use case narratives — the flow of events, as the UML people would say — are all too often missing the very essence of the use case, because behind every use case are business rules at work.

Failing to capture and verify the business rules which underlie the use case models can lead to the delivery of a software product which fails to meet the business needs. Formalizing business-rule capture concurrently with use case model development strengthens the delivered product through the synergy between use case models and business rules.

Business rules come in many forms. I think of them as terms, facts, factor clauses and action clauses, but there is no agreement on a standard taxonomy or set of categories for business rules nor should there be. The taxonomy should fit the problem. Some things in the "problem space" are more business-rule-based (e.g., underwriting, claim adjudication, financial-risk analysis, medical instruments monitoring). Other problems are more business-rule-constrained (payroll, expense tracking, ordering, etc.). This requires the selection or tailoring of a taxonomy and an accompanying business-rule template for any given business problem. The template provides a standard syntax for writing business rules in natural language. Such tailoring is beneficial since it requires us to understand the problem in greater depth by working directly with our business customers to perform this tailoring and to derive an appropriate business rule template.

Business rules can be linked to a use case or to steps within a use case. Business rules also can be linked to other models, depending on the depth of requirements traceability and the importance of specific attributes, such as source, verification process, documents, owner or risk. These other models can include glossaries, lifecycle models, activity models and class models. The business rules are thus reusable across multiple types of functional requirements and can be traced along with other requirements.

Besides use cases, other models can be useful for identifying and capturing business rules. For example, lifecycle models, such as a simple state-chart diagram, can be quite usable and understandable to business analysts, especially for problems in which a core business domain has many complex states. Even object-class or data models can be excellent starting points for problems that require understanding the domain structure, but which do not require a lot of "action" to take place.

Collaborate with Customers

Business-rule discovery and validation requires knowledge of the thinking processes of the business experts from whom we elicit functional requirements. Collaborative work in groups that include customers is most effective in the early lifecycle phases of planning and requirements analysis, as well as for ongoing project process improvement. Such collaborative work patterns can be used effectively to model business rules and derive higher quality requirements, which include business-rule requirements.

In the collaborative approach to business rules that I prefer, requirements analysts and business customers collaborate to create a business-rule template that is expressed in natural language, based on common sense and directly relevant to the business customer. In modeling use cases and user interfaces, business rules are explicitly derived and captured using this natural language template. After all, what makes sense to our customers and users is what we express in their own language.

Collaborative modeling of business rules can be a powerful, eye-opening process. It is amazing to see customers realize that their own business rules are unclear, even to them. Often they come to realize the rules are not standardized and therefore may be inefficient or even risky. Considering regulatory exposure and potential for lawsuits, collaborative modeling can convincingly make the case for immediate clarification of business rules.

Eliciting business rules can be a real challenge, especially when business experts do not agree on the business rules or when the business rules are unknown or very complex. I find that facilitated requirement workshops in which business rules are explicitly discovered, capture and tested within the context of modeling other things, such as use cases, is the most direct and productive tool for eliciting and validating the rules of the business. These workshops are planned collaborative events in which participants deliver products in a short, concentrated period of time led by a neutral facilitator, whose role is to help the group manage the process. Prior to the workshops itself, the participants have agreed upon what will be delivered and the ground rules for group behavior. The workshop process exploits the power of diverse people joined together for a common goal.

A successful requirements workshop requires considerable pre-workshop effort. Such collaborative events, when woven throughout requirements analysis, tend to increase speed, promote mutual learning, and enhance the quality of the requirements themselves. In workshops, or even in interviews with business experts, if the problem domain is very much 'rules based' (vs. rule constrained as mentioned earlier), using cognitive patterns is extremely helpful to accelerate the group's ability to express the rules. Such patterns, with their roots in knowledge engineering, model business expert's thinking process and enable the rules to emerge.

Some customers may wish to take on a business-rules approach because they want to uncover the real "dirt" of the business. They are ready to ask the "why" of the business rules. Actually, the question of larger purpose should be asked of any functional requirement. If the answer does not map to a business goal, objective or tactic, then the functional requirement is extraneous. Business rules exist only to support the goals of the business. Whereas good use cases represent the goals of external users or "actors," the business rules behind use cases are inextricably linked to the goals of the business itself. If not, the rules are extraneous and may even be in conflict with business goals. Thus, mapping business rules to business goals is a key step in validation and promotes traceability back from business-rule requirements to the business goals.

The Business-Rule Cure

Business rules are an ounce of prevention. Unless we get them, get them right and get them early, we are destined for problems in our projects and products. The project problems stemming from incomplete, erroneous or missing business rules include redefining requirements and re-testing results. The product problems are worse because, if the rules are wrong, in conflict or redundant, the users of the software product suffer. Unless we get to the very heart of the business requirements analysis, with the business rule written in text by business people themselves, we are doomed to passing incomplete, inconsistent and conflicting business-goal requirements ahead to production. Thus, we must get to the very essence of requirements with business rules, written by business people.

To get to the heart of the matter, active business sponsorship is absolutely required. The process can be acutely uncomfortable because business-rule capture and validation exposes the "undiscussables" — the unclear and conflicting business policies and rules. It also begs for rethinking the sub-optimal rules and requires the realignment of the business rules with the business goals. To resolve such issues requires business sponsorship and leadership. To go forward with a business-rules approach in the absence of such sponsorship is treading on very thin ice. Only the collaborative efforts of IT professionals and their business partners with the active involvement of business management can yield the benefits of a business-rules approach in quickly and succinctly cutting to the core of the functional requirements.

3.5.6 "Learning the Laws of Usability"

by Lucy Lockwood

Careful attention to the user interface is a hallmark of high-quality software systems. A well-designed user interface makes the user's interaction with the software more efficient, less prone to error, faster to learn, and more productive. Good user interface design goes beyond detailed consideration of the screen layout; it focuses on creating an effective architecture for the user's interaction with the system based on a thorough understanding of the work you need to accomplish. These guidelines should help you and your team improve the quality of the project's user interface design.

1. Design for real work.

Software is an enabling tool, letting us do things faster, more efficiently, and expanding our abilities in the world. Except for a few obsessed techies and dedicated quality assurance staff, most people don't use software for the sheer joy of seeing someone's code in action. Software is a tool, a means to an end, whether it is looking up tomorrow's weather on the Internet, heating up soup in a microwave oven, or running medical diagnostics on a patient. With this in mind, it behooves us as designers and developers to understand what our users want to accomplish and what they need the software to provide in order to succeed in their tasks.

Keep your focus on the actual work people need to do and consider why a user takes each separate action in completing a task. Beware of getting mislead by management or others removed from the actual work. Avoid getting seduced by the technology, too. It is easy to get carried away designing cool new features that solve non-existent problems and only serve to clutter and confuse the user interface.

Make the time and effort to do a thorough job of gathering information and analyzing the work and the needs of the software's users. Involve users in this process, but don't blindly accept a list of functional requirements thrust at you by users or your clients. After all, the users may be experts on their own work, but they are not experienced software designers and analysts. Ensure that your development process incorporates an effective method for collecting, organizing, and verifying information on the work to be supported. Educate yourself by learning the domain language, learning the work process, and understanding each requirement—including how and why it fits into the whole picture. Then incorporate the results of your analysis into the interface design process and use this knowledge to drive the overall software design.

2. Don't get concrete too quickly.

Developers tend to be solution-oriented individuals; we like solving problems and we like to see the results of our efforts quickly. At times our proclivity for fast results leads us to mediocre solutions. Modern visual development tools encourage such tendencies, permitting us to "solve" user interface design issues by simply dragging and dropping pre-made widgets onto our screens. Often, little consideration is given to whether a particular selection would best be handled by a dropdown picklist or by a combo box, for example. You will discover that you and your team create much more usable designs and develop superior solutions to your user interface design challenges if you resist the urge to get concrete early on in the development process. Hold off on deciding the implementation specifics until you have gained a better understanding of the work to be supported and have a good sense of users' intentions at each step in the work process. If you find your team jumping ahead to specific solutions, put those ideas in a "feed-forward bin" to draw on later in the design process.

Techniques such as abstract prototyping, which represents necessary user interface elements in a generic form by sticky notes, can help by letting your team focus on the overall user interface architecture first. Rather than getting caught up early on in bitmap design or widget selection, abstract prototyping lets you ensure that the software will provide all the necessary components to support the tasks and that you have arranged these needed tools and materials logically.

3. Avoid innovation for innovation's sake, but don't be a slave to fashion.

New is not necessarily better in interface design. New widgets and interaction methods can be powerful tools and a marketing advantage if well done. Those that truly advance the field will soon be copied by others and become part of our standard design vocabulary. However, much innovation in user interface design ends up providing worse usability than the traditional widget or interaction that it replaces. To make sure that users will understand your innovation and find it useful rather than confusing or awkward, involve user interface design experts in the process and perform careful usability testing to confirm the design's effectiveness.

On a related note, avoid creating art for art's sake. Some of the most unusable software I have encountered has been the product of graphic artists or would-be graphic artists who got carried away with the visual effects and ended up with controls that were unrecognizable, layouts that were unreadable, and navigation that was unworkable. Efficiency and aesthetics don't have to be mutually exclusive; indeed, the best software combines both. Recognize the limits of your artistic talents, however, and focus on creating effective, highly usable software that supports the intended uses.

Finally, don't assume your design has to slavishly copy a particular Microsoft format or use at least one of all the latest user interface widgets. User interface design does have its own fashion trends and staying current is important. New widgets and interaction idioms (as Alan Cooper calls them) are being created every day. Just as in houses, cars, or clothes, software styles change and evolve; you can often date a system's design simply by looking at the user interface style. While you don't want to create software that users immediately view as being circa 1990, you also don't want to go to the other extreme and get overly caught up in following all the latest fashions. When tabbed dialogs first appeared, I saw systems that used them everywhere, even in situations where they were awkward and inappropriate.

A safe bet for many developers is simply to copy whatever Microsoft does in design. There are many situations in which this is a reasonable choice, particularly if the software being created will be used closely with or intermixed with heavy use of Microsoft applications. Playing follow the leader is not a universal solution, however. In one memorable example, I helped lead a usability inspection of a currency trading system in which the user interface had been modeled on Microsoft's Excel. This design decision made it very difficult for the users to visually parse the information and led to awkward, inefficient navigation. What the currency traders needed from their new trading system was not the same user interface design that Microsoft had provided for their spreadsheets.

4. Strive for efficient interaction.

No one likes tedious, awkward work. Make sure your software doesn't add to the unpleasantness of someone's job. Count the keystrokes and mouse clicks needed to accomplish common tasks; make simple things simple, frequent interactions fast. Don't bury key information or capability four dialogs down or hidden in a mis-named menu item. Use modeling techniques such as navigation maps to help design for efficient workflow and to avoid long or dead-end navigation paths. User interface design metrics such as essential efficiency, task concordance, and task visibility can be used once you have a paper prototype and will quickly highlight any problem areas.

As one memorable Dilbert cartoon commented, some software designers inflict pointless extra steps on the user, punish them for any mistake, and essentially make them jump through hoops to get their daily work done. If some of your team members appear to hate users, bring in a therapist — but don't let them put the most common function under "Other" on the Edit menu.

5. Try out the interface for real work.

As a final measure of user interface design quality, have each member of the design and development team use the software for real work, at least a continuous half hour and a full hour is preferable. Use real data — the same data your users will be working with. If confidentiality or security concerns preclude using the actual working data, have the client or users create ample "pseudo data" (not a mere 10 or 20 records) that accurately reflects the complexities, errors, and subtleties of the live data. As a separate check, make sure you max out all the data fields in terms of input length and correct masking and filtering. I am always amazed at the frequency with which I encounter data entry fields that are correctly defined in the database itself, but which lack a sufficiently long, or properly defined, data entry field on the user interface. Try inputting twenty-five characters if your field length is such. Can you see them all on the screen? Does the zip code entry box handle foreign alphanumeric post codes if needed?

Yes, in many companies much of this should fall to quality assurance and to database design personnel, but surprisingly often it slips through the cracks. So, use the experience of trying out the user interface to test for it yourself. Consider this testing process to be like the auto assembly line final check — this is what the customer will see and experience. If the interface involves data entry, input 50 or 100 records, not just one or two. If the application will be used under odd lighting or noise conditions, try it out in such an environment. If rapid entry or rapid response to on-screen information is required, put yourself and your teammates under the same clock pressure. Then ask yourselves, "Would I want to use this software every day?" I hope these design tips help you and your team to answer, "Yes."

3.5.7 "Your Passport to Proper Internationalization"

by Benson I. Margulies

You've probably heard horror stories about how hard it is to modify software to work in Japanese or other Asian languages. On the other hand, perhaps you've encountered claims that such a process is a simple matter of extracting strings and translating them. In fact, there are many potential procedural and technical pitfalls in building international software, and there is no substitute for thinking carefully about the problem and formulating a plan of attack.

How Not To Internationalize

Here's a cautionary tale about a company — we'll call it Acme Products — that did not take an organized approach to internationalization.

The trouble began when the company decided to enter the Asian marketplace. Acme hired a new vice president for Asia, Clive, who in turn hired a new sales and marketing staff. Clive's team members spent all their time and energy firming up distribution relationships in Japan. They made no attempt to verify that the engineering was being done to provide them a product to deliver in Japan. It wasn't long before they had closed the first big deal.

Clive arrived at the corporate office about a month before the next major release of Acme's product was due to be deployed — just after feature freeze. Not only was making changes to the new release unthinkable, but all of the developers were already working flat-out on the release. They had no time to discuss a Japanese version of the previous release, let alone build one. Clive had no choice but to bring in new people to build the Japanese version. With no access to the over-occupied original architects or developers, they used the previous release as a black box in their attempt to fulfill their mandate: "Make it work in Japan very quickly."

From Multibyte Coding System character set nightmares to tricky Western SQL database support, from GUIs that assume Latin font metrics to poorly planned parallel development, myriad mishaps can occur in what looks like a simple localization exercise.

Clive wasn't a technical person. As far as he knew, this was a "localization" problem — all the English strings in the product needed to be in Japanese. No big deal.

Given that direction, the team set out to do a strict localization: Find the strings and translate them, and nothing else. They didn't anticipate what, if any, problems the code might exhibit when faced with Japanese text, and they didn't consider any new features.

Acme's development manager, Sam, didn't want to hear about this project. It had come out of nowhere, with a budget that was not under his control. He told his people to make a source snapshot of the product and "throw it over the wall" to the "outsiders" on the Japanese project. Sam's job was done.

Acme's programming shop was not accustomed to parallel software development on any scale. The release manager, Rhonda, was in charge of source control in conjunction with her release group. They used a source-management system with a single line of development from release to release and handled patches somewhat informally. They weren't prepared to handle an outside team making significant changes to the code in parallel with their ongoing development and maintenance process. To set up a real parallel development branch would have been quite costly the first time — especially since the Japanese project was off-site. Since this was neither in the budget nor the plan, no one in Rhonda's group wanted to hear about it. Rhonda ignored the Japanese project.

The intrepid Japanese development team got down to work anyway. Their main problem was strings — or so they thought. They had heard somewhere that message catalogs or resources were the "right" way to handle these, but catalogs or resources looked like too much work for the time they had allocated. The code was in C and had constructs such as:
```
static char *strx[] = "A typical string: our Hovercraft is Full of Eels";
```
The compiler wouldn't let them replace that string with a function call. Other strings were in performance-sensitive spots where the cost of a function call would be deadly. They simply didn't have time to deal with the changes required, so they made a fateful decision: They built scripts to replace the strings in-line, creating a mutant version of the code base with all the English replaced by Japanese.

In order to put Japanese strings into the source, they had to get some Japanese strings. That wasn't too hard. Plenty of folks "know Japanese." The team didn't consider whether those people were qualified to translate technical terms. After all, they were in a hurry, and one of the team member's boyfriends could knock off the work in the evenings in no time flat.

Once the translation was done, the developers stuck it into the code without further thought. No one edited it.

The Acme Japanese development team started having problems with testing. To begin with, Acme had few, if any, formalized testing procedures. There was a quality assurance team, and the members of that team had a body of folk wisdom for their release testing. There were some automated scripts, but these all assumed English text in the user interface. Plus, the scripts were implemented in a testing tool that didn't support Japanese. The team struggled with the difficulty of obtaining, installing and maintaining localized Japanese versions of the operating systems for testing.

Finally, they delivered the first beta version to Japan. Soon after, the defect reports began to arrive. The team had expected some minor visual blemishes and idiomatic inaccuracies. They hadn't expected the following:

Storage corruption.   Pointers and other data were corrupted due to buffer overflow.

Ungainly dialog boxes.   Acme's product had a Windows GUI, which, like many GUIs, had a population of dialog boxes. Some were carefully designed to fit into 800x600 pixels. Imagine the team's surprise when they discovered that the dialog boxes didn't fit into 800x600 on Japanese Windows, even without any Japanese text!

Mysterious database errors.   The application worked with a SQL database. Down in the depths of the product, the code set is specified by an obscure parameter that determined the character encoding for communications with the database. In this product, the parameter depended on a default setting that didn't support Japanese.

The Japanese team had little database expertise, and Acme's database experts were far too busy with the new release to be bothered. Not only didn't the team know about the environmental parameter that controlled the character encoding used on the client, they didn't know that there was a critical parameter that had to be set when the database was installed and configured. As a result, they went along using a "Western European" database and could not understand the strange misbehaviors and errors that resulted when they tried to store Japanese text in some fields. Furthermore, they just didn't understand why fields that were always long enough in English weren't long enough in Japanese.

Poor translation.   The translators weren't qualified, and no one had checked their work.

Missing features.   The product collected first and last names, but the developers had not made provisions for the pronounceable versions of Japanese names.

Against the odds, the team eventually delivered a product. It was late, it was missing critical features, and it had bugs. (Other than that, it was fine.) When mediocre sales followed, the top managers all blamed each other. Clive, the Asia vice president, blamed Sam and Rhonda in development and release management for failing to support his effort, while Sam and Rhonda felt that the mediocre sales proved what they had been saying all along — there is no real money to be made in Asia.

The Japanese development team tried to merge their work into the source for the next release, but without a code branch in a source management tree (a basic capability of a good source management tool), this was impossible. In any case, the in-line Japanese strings couldn't be checked in replacing the English strings, and the Japanese version was now in permanent limbo. Japanese customers who encountered defects had to wait for someone to manually port the fix from the main English code base to the Japanese version.

In spite of the difficulties, Clive did not give up. He doggedly scoured for sales opportunities. And he found them. However, he had a huge problem — the new customers required features from the new release. And some of them were in China instead of Japan. Once again, the weary team suited up, resigned to starting the entire process over again. And here we leave them, slogging along and cursing their fate, the perpetual outsiders. The developers hate them, and they aren't too popular with Clive either because of the product's poor performance. Who would want to work on a team like this one?

Details of MBCS Issues

Acme's code was full of little constructs that were all set to misbehave when presented with text in a Japanese character encoding. Japanese text is usually encoded in a Multi-Byte Coding System (MBCS). That means that some characters are one byte long and others are two. (On certain platforms, some characters even require more than two.) Many common coding clichés don't work with MBCS text. Code like

```
for(sp = buff; *sp; sp++) { … }
```

is not processing characters, it's processing bytes. If it stops in the middle of a character, or compares a single byte to some ordinary ASCII character, it can end up in the middle of a multi-byte character, with disastrous consequences. For example,

```
strchr(buff, '/')
```

can end up returning a pointer into the middle of a character.

What Went Wrong?

The Acme project suffered from a series of problems, and each problem fed the next. That doesn't mean that you can't get into trouble by only following parts of their example; you can get into big trouble with any one of them. To avoid internationalization problems, consider your organizational structure, process and technical requirements. Here are some of the specific mistakes in the Acme story:

Bad organizational structure.   You can't get a good result from internationalization if you don't treat it as part of your core business. If you ghettoize your international initiatives, they will suffer from the lack of communication, coordination and buy-in. It's a good idea to have a separate marketing and sales vice president focused on the specific regional and cultural issues.

Failure to coordinate.   Because the internationalization process wasn't integrated into the overall development plan, the Asian requirements — which merited the initial attention of the architects and developers — weren't included in the release.

Failure to understand the requirements.   As with many other foreign markets, Japan has unique needs. For example, Acme's product required support for pronounceable Japanese strings, called furigana. These must, in some cases, be displayed atop a string in a typesetting convention called ruby text. The product also needed support for the Input Method Editor that allows the input of ideographic characters, as well as support for the Japanese rules for breaking lines. Since no one took the time to analyze these potential requirements, they were all left out. The first mistake was disastrous: Without *furigana*, Acme's customer service representatives couldn't pronounce the customers' names.

Dialog Box Woes

The font metrics are different in Asia. If you carefully design a dialog box to fit in 800x600, it won't fit as soon as you try to run it on an Asian system.

Individual strings may be bigger (overflowing their graphical boundaries) or smaller (leaving ugly white space).

Your layout may embed an English grammatical assumption. The right word order for English is not necessarily the right word order anywhere else. For example, From [EDIT CONTROL] To [EDIT CONTROL] is an ordering of screen text that assumes English grammar; in Korean, the user fields would not fall in that sequence.

In Asian fixed-width fonts, characters come in two sizes: full-width and half-width. The ideographic characters are full-width, the Latin characters are half-width. And then there are special full-width Latin characters. If you have code that sets up neat columns by assuming a matrix of fixed-width characters, it won't look good until you teach it to count on halves of fingers.

Failure to appreciate the technical challenges.   Even taking English software to Europe can yield snags. Acme had a much harder problem in aiming for Japan. Here are a few of the biggest obstacles they slammed into:

- Their GUIs embodied English grammar and needed significant rearrangement.
- The GUIs assumed Latin font metrics and thus didn't fit on the screen in Asia.
- In Asian languages, text is represented in the Multibyte Coding System (MBCS). Acme's code used some of the common clichés for strings that corrupt MBCS text. See the sidebar for some gory details.

- Acme's product used a database and foundered on the complexities of the SQL database's international support.
- The code sorted strings by pure numerical order.
- The code used third-party components that didn't work for international text.

The code included latent defects, especially storage defects, that are common to C or C++ code. Without excellent test coverage or a good storage defect detection product, these problems lie low. Heaps are surprisingly tolerant of some mistakes, but changing the lengths of strings is a good way to change the allocation pattern and convert latent defects into blatant defects.

In short, there is no such thing as "localizing" a body of code that has never been internationalized before. This is always an internationalization project. It might be a small one or a large one, but it is never absent. Once the code has been internationalized and the changes integrated into the code base, then there is the possibility of pure localization to adapt it to additional countries.

Failure to account for parallel development. The team made no allowances for proper source management during parallel development. Internationalization projects are textbook examples of the importance of source management. They involve many small changes to a large body of code, and they must often occur simultaneously with other development efforts.

The international team has to understand the development process, including the flow of requirements into design and the schedule for releases.

Weak translation and editing. Acme is something of an extreme case. Many companies setting out to localize at least manage to hire a professional translator. However, even a professional translator does not guarantee success. What was missing from this picture? Editing. To speak in bald economic terms, translators are often paid by the word. They have an incentive to rush. And, like the rest of us, they make mistakes. Translators are a species of writer, and any good writer knows the adage, "He who proofs his own copy has a fool for an editor."

Leaving strings in place. The quickest way to localize code is to take the source and replace the English strings with strings in another language. It may be quick, but it essentially dooms any attempt to maintain a single source that works in multiple countries.

Testing. Acme's problems with testing were like Acme's problems with source management. A relatively informal development process that worked "well enough" for one country didn't work well at all for a multilingual product. Acme wasn't prepared to have international testers show up and do productive work.

A Spoonful of Process

To succeed with internationalization, modify your development process to accommodate the effort. You'll find that internationalization turns up at all phases of the project. This may sound daunting, but keep in mind that a small amount of effort and attention early on will save you a lot of work down the line.

Successful internationalization starts with good communication among all the participants. If there is a specific international sales and marketing team, its management should establish strong lines of communication to the development management team. The international team has to understand the development process, including the flow of requirements into design and the schedule for releases. A successful international initiative must be launched in the entire company, not just in an international group, subsidiary or division. It has to be sold to everyone, and the budget has to make realistic provisions for everyone's efforts.

Ideally, internationalization should be the responsibility of the core development group. They know the code best. If international support is a full-fledged requirement for a normal release, then the developers can ensure that the required changes are integrated into the architecture. This is an important part of giving the entire enterprise an international outlook. Even if the core developers can't do the work, it is a good idea for their managers to have some oversight over the internationalization developers.

If, for reasons of budget or schedule, internationalization has to live outside of the core development organization, it is especially important to foster strong communications between the international sales and marketing team, the international development team and the core development team.

Requirements

If you want your code to support international deployment, say so at the very beginning. It is much easier to build U.S.-only code than international code. Schedule-pressed developers will never allow for international support if you don't make it part of the requirements. Don't listen if someone tells you that international support is zero-cost if you only take it into account at the outset. That's not true. It will cost something. Keeping it in mind from the outset, however, will make it cost much less.

Technology

Internationalization technology is a broad topic. Almost anything that touches text can turn out to have or create internationalization problems. Here is a survey of the major difficulties that many projects encounter.

Third-party components. The most dangerous aspect of any code base, from an international standpoint, is the presence of third-party components. If the component doesn't work right for international text, you have to either get the vendor to fix it, find another vendor or recreate the functionality for yourself. Any of these can have a long lead-time. Survey your third-party components early, and find a strategy for them.

Strings. Most likely, your code has strings seen by human beings. You have to find them. You have to distinguish them from strings that are not seen by human beings. You have to arrange for them to be translated in each release. And, most importantly, you have to make sure that ongoing development does not undo all the work of finding the strings and making them available for translation.

MBCS. The classic bugbear of Asian internationalization is MBCS. These days, most programmers believe that writing 8-bit clean code is a good idea. Characters that occupy more than one byte, and especially variable numbers of bytes, are another story. Almost all code that processes text is unsafe for MBCS text and requires remediation.

Unicode. Unicode, also known as ISO-10646, is a standard character set for multi-lingual text. The Unicode Standard defines a very large character set and several character encodings. All commonly used languages' characters are represented in the character set. Unicode provides a uniform and organized approach to character properties and managing scripts that run right to left.

The character encodings offer you a choice of representations. One choice is UCS-2. In this encoding, all the characters are same size in memory: 16 bits. In UCS-2, you don't have to worry about splitting a character in two. You don't have to worry about the number of elements of the array versus the number of characters. You do have to modify all your code to declare textual data to be represented as 16-bit items instead of 8-bit items. This can be a very expensive conversion to existing code.

To avoid this conversion cost, Unicode defines several alternative character encodings called "transformation formats." The important ones are UTF-8 and UTF-8-EBCDIC. Both of these are MBCS formats, and thus subject to MBCS coding problems.

While the transformation formats are MBCS, they are "friendly" MBCS. Simple ISO-646 (ASCII) characters never occur as a part of multibyte characters. Therefore, many common operations on strings (for example, strlen in C) are valid with UTF-8.

The GUI. Even simple GUIs can encounter serious problems. For an example, consider dialog boxes in Windows. (See Dialog Box Woes for some of the gory details.)

Asian languages have many more characters than fit onto any imaginable keyboard. To allow input, systems provide Input Method Editors. These allow the end-user to construct characters by typing phonetic or other representations of them. If you use standard GUI components, you will find that the IME "just works." If you do your own keystroke handling, you may need some serious effort to coexist with the IME.

Databases. The major relational databases handle the major international character encodings. However, you may have to make complex or troublesome changes to make use of their support.

Printing. If your product has printer drivers, it is very likely that they will need significant enhancement to print Asian text, even in common printer languages such as PCL and PostScript.

— Benson Margulies

Internal requirements take two major forms: the language-neutral requirements and the specific target requirements. Examples of language-neutral requirements are:

- The code shall support localization with few or no modifications to executable code. This includes dialog layouts.
- The code shall operate properly with MBCS international text. (Or with Unicode, another way of dealing with MBCS languages).
- The code shall operate correctly with text in several different languages.
- The code shall use local-sensitive sorting, currency formatting and date formatting.
- The second flavor is specific requirements for specific target markets. Here are some examples:
- Japanese *furigana*, as previously described.

- Chinese elaborate numbers. In addition to Latin digits, the Chinese script includes a set of elaborate ideographic characters for numbers. These are used on financial documents in much the same way the spelled-out numbers are used on checks in English.
- Line breaking. For each Asian language, there is a set of rules for inserting line breaks into text. These have nothing to do with word boundaries.
- Dates and calendars. Japan, Korea and the Moslem world all have alternative calendars that are in common use. Depending on the context in which you are presenting or soliciting a date, you may have to work with an alternative calendar.
- Numbers that identify people. The US has nnn-nn-nnn social security numbers. Other countries have a wide variety of identification schemes with varying formats.

Parallel Development

In an ideal world, your development team would already include a few international aces, thus obviating the need for a separate group of experts. But perhaps your company has international developers in a special group or prefers to outsource the work. Acme took that difficulty and exacerbated it even further. To work successfully with parallel development teams, you need good communication and strong source control.

Consultants

Internationalization projects often involve outsourcing. Specialized expertise is needed, often in conjunction with hurry-up schedules. Some consultants will educate your core developers as part of the process. Some won't. Stick with the first kind.

Once your code is internationalized, it has to be localized. That is the process of producing translations of strings and other materials for each target. Unless you work for an enterprise that is so vast that it can afford to have a staff of professional translators, use an outside vendor for localization. Your international developers (in-house or outsourced) have the job of making sure that the necessary materials are easily identified, handed to the localization vendor and reintegrated after translation.

Architecture

Once the requirements are set, the next step is architecture. There are entire books — Ken Lunde's *CJKV Information Processing* (O'Reilly & Associates, 1999) is one — that discuss the various alternatives, so I'll restrict myself to a single example. The most basic question is where to put the strings. In many cases, there are three alternatives: Windows resource files or Java resource bundles, some other sort of message catalog file or a database.

When deciding among these, it's important to question whether you need to access strings from more than one language at a time. If you are implementing a Web server, for example, you may need to grab the strings that apply to a particular client users, since different clients are in different languages. Some message catalog systems enforce a single, static language selection. One of those systems would be a very poor choice in a Web server. On some operating systems, the run-time library has a single language setting (the locale) for an entire process, and there is no thread-safe way to change it for the life of the process. On such a system, you must either ensure that a particular process handles only requests for a particular language, or you have to substitute a thread-safe mechanism for the system mechanism.

Another important consideration is performance. If you use a database to store strings, beware of introducing a database query latency into a performance-sensitive code path. One good strategy is to store a time stamp in the database that records the last time that the strings were updated. Clients can then maintain a local cache of strings and only retrieve them from the database when they change.

You can pseudo-translate to find and flush out many defects before the translations are available.

Schedule

Internationalization first shows up in the schedule as the features chosen in the architecture are developed. So far, we have simply added more development tasks to the schedule. Later on, things get more complicated, and many of these things are the actual localization of the code.

Translation and localization take time. Before you can start translating, you have to have a set of strings to translate. You have to freeze the strings. If you can't freeze them altogether, you have to start to track the changes to strings, so that you can send incremental jobs to the translator(s) to keep up with developments. One way to expedite translation is to start with a glossary. Extract a list of important words and phrases from your code and documentation. Send it off to your best (and perhaps, most expensive) translator. Have it reviewed by an editor. When the time comes to send the bulk of the text off for translation, have the translator work from the glossary.

Quality

Once you send strings off to the translators, it takes them some time to complete the work. In the meantime, you may be wondering how well all your internationalization changes have worked. "Gee," you may think, "too bad we can't start testing this stuff yet." Well, you can. You can pseudo-translate to find and flush out many defects before the translations are available.

In pseudo-translation, you pretend to localize the product. You take the original translatable materials, and you decorate each string with (for example) a few Chinese characters. Then you test the product, looking to make sure that the Chinese characters appear, correctly, in all the right places.

Of course, you must leave room in the schedule for late translation fixes. In spite of the best efforts of translators and editors, mistakes and misinterpretations will turn up in the late stages of testing. Allow time to send them off for corrections.

A Business Imperative

Now that you know the truth about internationalization — it isn't easy, it isn't simply a localization project, and you can't accomplish it on a shoestring — you can focus on the rewards of successfully translating your software for use in other countries. International Data Corporation estimates that worldwide business-to-business e-commerce will grow to $30 billion by 2001, while by 2002, non-English speakers will make up more than 50 percent of the world's online population.

With more than half of the world's Internet users predicted to be non-native English speakers by 2002, going global is not merely a business advantage in the 21st century; it is a business imperative.

Using Unicode to get you there is the most efficient and promising way to ensure your worldwide engineering process is effective, affordable and rewarding.

3.5.8 "Thirteen Steps to a Successful System Demo"

by Larry Runge

Follows these tips and give system demos that will impress
even the most nontechnical audience.

Good software development is as much art as it is science, although many business managers may consider it to be a black art or sorcery. A skill that is every bit as important as developing an effective user interface is the art of giving an effective demonstration of the software. I have seen excellent, innovative business applications fail to gain the notice they deserved because of ineffective demonstrations.

A good demo can sell your system, and it also can sell you and your abilities to upper management or to potential customers. Good demos are not that difficult to perform, as long as you follow a few guidelines. Here are 13 steps to producing an effective system demo.

Step One: Budget Your Time

Determine the total amount of time you have to do the demo. Determine if this includes time for questions, either at the end of or during the presentation. It's very important to adhere to your allotted time; the entire demo must be planned and structured to fit within your given time frame. If you run out of time, you may miss the chance to demonstrate valuable aspects of the new system, but if you rush in an attempt to fit everything in, you run the risk of alienating the people you are trying to impress.

Step Two: Make a List

Make a list of the key points you wish to address during the demo. These need to be from the user's perspective, not from a technical perspective. To do this, list the features the system provides, then rephrase these features in terms that describe the business benefits they bring. For example, perhaps the new system allows the user to perform functions with a single screen that in the old system required two or more screens. Thus, from a business standpoint, the feature may cut in half the time required to perform the process in the old system.

Features are things that the applications does or has. Benefits are things the customer will get by using the features. For instance, in a flight simulator program, all the instruments — the altimeter, the altitude- directions indicator, and so on — are all features. The benefits of these are that customers will improve their ability to fly by instruments if they use them. Thus, in a demo of the flight simulator, you would talk about how the features of the program would improve the customer's ability to fly by instrument.

Step Three: Prioritize Your Points

Key points are different than features. Again, features are part of the software. Key points are important statements you wish to make during the demo. For example, key points for the flight simulator program might be:

- It will run on PCs and Macintoshes.
- It offers airport landing strips for 50 different cities.
- It is priced the same as two months of cable TV.

You can prioritize your key points two basic ways. The first is the "newspaper" or "inverted pyramid" structure. In a newspaper story, reporters put the most important facts first and follow with less important facts. The second is the "Hollywood" or "crescendo structure." In this approach, you save the most important point for last and carefully build increasing levels of excitement as you progress.

The first approach is best when you're concerned that the demo may be cut short due to time constraints. The second is ideal when time is less of an issue, and you really want to make an impression on the people attending the demonstration. Regardless of which approach you use, ask yourself the following questions:

- Can you combine one or more points to be more effective?
- Does one point fit into or set the stage for another?
- Are there natural segues among the features?
- Does one key point reinforce another?

Step Four: Determine Key Features

Determine what features best support the key points you want to make. Remember, for an effective demonstration, all key points should be demonstrated and reinforced using features the system provides. At this point, you should have a list of key points you wish to make, prioritized in order from most important to least important, with one or more systems features written down next to each of your key points. This is now the outline you will use to script your demo.

Step Five: Develop a Script

Develop a script for your demo. Just like a script written for a play, television show, or movie, your script should have a beginning, middle, and end. The beginning introduces your audience to what the system is supposed to accomplish. The middle will demonstrate the features that cover your key points and should roughly follow the outline you developed in the previous step. The end is where you summarize the benefits and wrap up the demo. This is very important, for nothing is more embarrassing than to just trail off at the end of a demo without tying everything up.

Step Six: Add Detail

Add just enough detail to make a smooth demo — don't try to demo each and every minute feature or capability. The purpose of a demo is to demonstrate and convince — too much detail will lose the attention of your audience. Too little detail may cause important points to be missed, while too much may cause the eyes of your audience to glaze over. Let's say, for instance, that you're giving a demo of a word processor which offers 25 different fonts in 16 different colors. Just mentioning in passing that it can do this is too little detail — you've missed the chance to "wow" them. A better approach would be to show them a couple of the different fonts in a couple of different colors. However, if you insist on showing them each font and color, you will put your audience to sleep.

Step Seven: Make Brief Notes

Although you have a planned script, rough-out your supporting commentary. At this point, you should have short explanations for each feature. Don't write these down other than perhaps a key word or two to jog your memory, and don't try to memorize your script. Doing the latter often causes a demo to come off sounding mechanical. And never, ever write down your text and read from it for the demo — this only works for professional news commentators. You want your demo, regardless of how much preparation and rehearsal you've put into it, to come off sounding spontaneous and fresh.

Never forget that nontechnical people don't understand jargon — so don't use any unless you want to thoroughly confuse your audience. Use analogies whenever possible to explain complex technical features. When Thomas Edison was trying to get people to replace gas lamps with his new light bulbs, he used terms that everyone could understand. At that time, there were "gas mains," so Edison, instead of creating an entirely new word for his switch boxes, called them "electrical mains."

One hindrance to people adapting to computer technology is that they have to learn a new language to understand what we're talking about. Do your presentation in English, or French, or Japanese, or in any other language that your customers understand, but — at all costs — do not do it in techno-speak!

Step Eight: Make a Dry Run

Rehearse your demo three times. This is to get the tempo down. Time your third dry run. This will tell you approximately how long the demo will take. Don't worry about adding or cutting material at this point. The things you should be looking for now are:

- Does your presentation flow naturally?
- Does it wander from your key points or lose focus?
- Does it follow the structure you've selected?

Step Nine: Expand or Contract

Once you know the approximate time length of your demo, you can decide whether you need to reduce the amount of material you planned to cover or beef it up. In practice, most people tend to speak faster during an actual presentation than they do during a dry run, so you may need to make it a bit longer than your actual allotted time.

Step Ten: Prioritize the Demo

No matter how well-rehearsed you are, your presentation can still run over its time limit — especially if a lot of questions are asked during the demo. Prepare for this in advance by looking for segments you can skip if you find you're running short of time. This way, you have the option of skipping over less-important segments if need be, while still covering the most important features. Never run past your allotted time, unless you've been told to take as long as you like.

Step Eleven: End Effectively

Plan an effective ending for your demo. Don't let it trail off, as this dilutes the effectiveness of your presentation. Just as a Hollywood movie needs an effective ending, so does your demo. Remember, if you're using your demo to sell a project or technical approach, you need to close your sale, just as a professional sales person closes a sales pitch.

How do you do this? First, the ending has to come at a natural break, to keep from leaving the impression that there's more to come. Like a sidewalk leading to a doorway, your path to the conclusion must be obvious. Frequently, this mirrors the logical conclusion of a unit of work.

For example, a unit of work for a word processor program is typically

- Determine need for document or memo
- Decide upon content
- Use the word processor to create the document
- Use the word processor to print the document
- Distribute the document.

Thus, if you were demonstrating a word processor, the logical conclusion to your presentation would be when you print the document either to paper or to a fax board, for that is the last part that the word processor will play in this sequence of actions.

Finally, make sure you verbally indicate you're at the end of the demo. An example of this might be, "And, this concludes the demonstration of this wondrous application! Do you have any questions I can answer or any comments you'd like to make?" After any subsequent discussion ends, quietly leave the room. Unless you're told otherwise, never begin breaking down the demo equipment until the audience has left the room. Doing otherwise disrupts the impression you've created of a professional business meeting — remember, you're not a roadie at a rock concert, you're a professional.

Step Twelve: Dress Professionally

Dress professionally for the demo. It adds credibility to your presentation. Of course, it's possible to overdress, depending upon the circumstances. In formal demonstrations at professional conferences, it's customary for speakers and presenters to wear business attire. In a business environment, you should target your dress to match that of the most senior person who will be attending the demonstration. If the presentation is for the chief financial officer at a conservative company, business attire will again be appropriate. If the company has a casual dress policy, then that will be appropriate. Be aware, however, that there are differing degrees of "casual" so, again, plan your dress to be in line with that of the most senior person there.

Step Thirteen: Plan for the worst.

Frankly, I've seen very few demos where something unplanned did not occur. It's almost an adage that the most carefully prepared demo can wander awry. Sometimes these are things beyond your control, such as when you need an analog phone line for a modem, and the only lines going into the room are digital. However, many times problems are encountered by simply deviating from the script.

For instance, you may practice using the script, and everything works fine. But during the demo, you deviate from the well-rehearsed script, and then something blows up or fails to work. Usually, this is a problem that you would have encountered had you tried this particular action beforehand. I've seen this happen dozens of times. So, never deviate from the script to show something you hadn't planned, and always do a dry run just before the demo. To do otherwise is to beg for disaster.

What do you do if the system crashes? Grin and reboot it. Make a joke, if you can, and say that Murphy's Law will always get you. Then continue speaking while restarting the system, just as if nothing happened and summarize the benefits they have seen up to this point. When the system is ready, pick up the demo where you left off — unless you had wandered off the script when the crash occurred, in which case you should return to the script, and things should be fine.

Finally, expect to be asked a number of questions. Generally, these will relate to things you've shown and should be easy to answer. If the question relates to business activities instead of your software, defer these, if possible, to someone intimately familiar with that part of the business.

However, make sure you answer clearly, concisely, and in a language that the customer can understand. Do not give long, rambling answers to questions. Just as your code should be tight and efficient, so should your answers.

Never try to bluff your way through a question — a snowjob is as easily recognized as a tornado in a parking lot, and you'll lose credibility. Frankly, there's absolutely nothing wrong with saying, "I don't have the answer to that, but I'll get it and get back to you." No one person can always know everything, and true professionals are willing to admit this.

Try to anticipate questions ahead of time, and prepare answers for these if they're asked. Finally, during prototype reviews and development demos, the toughest question is likely to be, "When can I have this?" Always be extremely conservative when promising delivery dates.

Wrapping It Up

Giving a demo in a professional manner is every bit as important as coding the underlying applications correctly. With a little planning and preparation, and with these steps, you'll be surprised at how quickly your demos can become effective and professional — and how quickly people attending your demos will associate those same qualities with you.

3.5.9 "Real Life Requirements"

by Larry Constantine

The first Management Forum — Live! confirmed that understanding the difference between what users want and what they need is key to quality software development.

It's one thing to pound out a column with ample time for reflection and the opportunity to revise and edit your words in the bright light of the morning after. It is quite another thing to solve major problems in real time, sitting across the desk from an important client or in front of a large audience. That's precisely what we set out to do at Software Development '98 Conference and Expo in San Francisco.

Posing the Question

What are your biggest software development management problems today? We posed that question by e-mail to several hundred registered attendees for SD '98's Management Conference. Then we assembled a team of experts who write for this column and put them on the spot in *Software Development* magazine's first Management Forum — Live! Flanked by six-foot projected images of themselves from two video cameras, the panelists were challenged to come up with workable solutions and usable recommendations in real time. Facing a large and largely skeptical audience of software and applications development managers, the panelists acquitted themselves quite well.

It helped that the panelists represented a range of backgrounds and experience. They included Peter Coffee, a well-known columnist and advanced technologies analyst whose recent rant in these pages challenged us to eschew excuses; Jim Emery, an associate provost and professor of systems management at the U.S. Naval Post-Graduate School, who offered the bifocal perspective of an academic who has studied and taught software development approaches, but who has also had to honcho mission-critical projects to completion; Karl Wiegers, a contributing editor to Software Development, who has also been a process improvement leader at Kodak and has a Management Forum column on project management in the works; and Norm Kerth, who described himself as president, founder, and head janitor of a one-man consulting operation that provides people-oriented guidance in object technologies for clients like Nike. His column on learning from project failures will be in a forthcoming issue.

Everyone's management problems are just a little bit different, but we work under common constraints typical of the times and culture. From what attendees reported to be their biggest challenges, some broad themes emerged. As moderator of the panel, I presented these themes, along with a sampling of representative questions, then let the panelists take off at will. By the time we landed at the end of our allotted time, the panelists had filled more than an hour and a half with wit, wisdom, and solid advice. One column is hardly enough to cover such a wide-ranging discussion — and only the videotape does justice to Norm Kerth's inspired sleight-of-hand impromptu magic trick — but I thought a synopsis of comments on one of the topics might be worthwhile and useful.

Don't Squeeze

If the e-mail poll is any indication of the general state of affairs, perhaps the leading challenge for software managers is feeling squeezed between the demands for quality and the pressure to meet tighter and tighter deadlines. The pressure is increased by the slippery slope on which so many projects seem poised, resting on requirements that are hard to identify, difficult to understand, and keep expanding as the project progresses. Many development managers feel as if they're balanced on a California cliff soaked by the rains of El Niño, with muddy requirements and an uncertain foundation threatening to give way at any time, dumping the project and its perpetrators over the brink.

"Our biggest problem is meeting deadlines while keeping to quality standards," responded one manager. Another complained about "coping with feature creep, ad infinitum, ad nauseum." Yet another referred to the problem of scoping requirements to fit within schedule constraints. "How," lamented another attendee, "can we build software where the only consistent rule is that all rules are subject to change?" These are not new problems, but it seems the vise grip of shortened release cycles paired with lengthening lists of requirements is felt especially acutely by many of today's managers. Is this just the case that project reality bites or is there something we can do to cope?

Peter Coffee kicked off the discussion by pointing out the numerous ways teams can fail when it comes to requirements. Many projects crash and burn because the developers fail to define requirements with sufficient care and precision. Other projects fail because the developers *did* meticulously define requirements. Fixing requirements too rigidly, too early, or too thoroughly can be as much of a problem as giving them short shrift. A preoccupation with rigorous requirements can lead to "paving over the cow path," in which new software replicates all the warts and wandering workflow of manual systems or outmoded applications being replaced. Even if delivered on time, the resulting over-specified systems will not serve the needs and interests of users well. The trick for developers is to define enough to know what to start building but not so much as to cast the code in concrete overshoes. Full specifications may take so long that user needs have changed or passed altogether by the time the requirements document is complete.

Creep Show

Requirements are not, as Jim Emery noted, the final word on anything. All requests and requirements from users must be examined with a critical eye. In fact, one of the best ways to meet requirements successfully is not to meet them all. Most systems are best deployed with an initial set of features that satisfies the most critical requirements first, saving additional features for future refinements that can benefit from experience and the perfect perspective of hindsight.

Users and clients are better able to accept an early but scaled-down release when they know they can count on the development team to deliver quality regularly. Norm Kerth suggested the goal should be to establish a culture of commitment, with regular, reliable releases on a three- to six-month cycle. Not only is this model more timely, but it also gives developers a tool to manage requirements creep. Instead of expanding the scope of the current project, late breaking requirements are deferred to the next revision cycle, which clients know will also be completed on schedule because its scope will also be actively controlled. It may require several rounds of refinement and release before client confidence is sufficient, so developers need to take the long view of educating their customers.

Of course, it is not enough to understand requirements in a technical sense. Development managers also need to understand their customers' definitions of success. A cost-conscious client on a tight budget may not be impressed when you deliver extra features that double the capability of a system but put the project 10% over budget. And no client is likely to be content to take delivery today on an affordable system that meets last year's needs. Since our definitions of success are often different than the customer's, it is important to make both of these explicit when defining requirements.

Drivers Wanted

In understanding requirements, it is also vital to distinguish drivers from constraints, as Karl Wiegers pointed out. Drivers represent performance or functional objectives that are vital to the business success of the delivered system. Technical and resource constraints define the boundaries within which a project must be managed. Not everything can be either a driver or a constraint. Some facets of any project must be recognized as trade-offs or options. Full features and functionality can be delivered even under highly restricted time schedules, but only at the price of quality or at added cost in development resources. Quality and reliability can almost invariably be maintained, but these may sometimes require restricting the scope or relaxing the schedule.

If clients don't recognize trade-offs, then there are no project management solutions and the project ends up doomed to failure. Part of establishing requirements, therefore, is educating users and clients to the intrinsic trade-offs. If we fail to inform clients of the costs, added features will appear to be free. Offered a choice between a full-featured Lexus and a stripped-down, second-hand Yugo, most of us would take the Lexus if the cost were the same. When developers do nothing but nod and take notes following every request, clients never learn they cannot have it all.

Kerth described a game that has been used to help in this educational process, a kind of hands-on metaphor for the popular technique of quality function deployment. In the game, all the sundry features and functionality desired for a proposed system are marked on a collection of various wooden shapes. Users are then invited to fit the ones they want or need into a box that is too small to hold all of the shapes.

As Coffee pointed out, educating and negotiating with clients is complicated because we can never pin down the exact cost of an isolated function. It depends. Most important, it depends on the quality of the design and the robustness of the implementation. It may be possible to incorporate yet another function at a modest cost so long as we can tolerate a clumsy user interface or are willing to accept that future revisions or improvements will be expensive. A rugged, broadcast-quality video recorder can cost 10 times as much as a consumer-oriented VCR, even though both may perform the same functions. Customers, users, and our own upper management can certainly understand such trade-offs, but we must take the time to make sure they understand how these apply to software. An efficient and adaptable implementation of software will cost more to program than a jury-rigged hack that will crash whenever it is modified, but how many users appreciate the difference?

Wish Lists

As Emery added, no matter what clients tell us, it's important for analysts and project managers to reject what we know is wrong. Users often ask for things that are unrealistic or not in their best interests. Ultimately, we are in the business of delivering solutions to real problems, and that requires us to give clients what they need more than what they want or claim to want.

Even the way we pose our questions to users will shape what we finally face as so-called system requirements. When we simply ask users what they want, they will tell us something, whether they know what they want or not. If we ask them what else they might like, they will invariably answer again — and keep answering every single time we ask them. Unrealistic and overly ambitious requirements often arise from trying to please users too much or from inviting their requests too simplistically or too frequently. If we try to play genie and grant client wishes, we are apt to construct castles of code in the air — baroque applications, bloated with features that meet no real needs — with little hope of delivering on time or under budget.

Use Cases

Distinguishing needs from wants emerges as a key to managing requirements creep. Wiegers has found that use cases are a powerful technical tool that can go a long way toward drawing that distinction and offering developers some relief from the pressure to deliver everything. To be successful, a system doesn't need to satisfy every wish or fantasy of its users, but it must meet a core of critical needs. What is critical and what is merely decorative? The most important capabilities are those that let users accomplish work, making it easier, faster, and more valuable to the organization.

Analysts need to understand what users are trying to accomplish, and use cases can be a tool to aid in this task. A use case represents an external, black box view of the functionality delivered by a system to its users. To avoid unnecessary features, Wiegers recommends use cases focused on the purpose of interaction rather than the mechanics, the very kind of essential model we developed to enhance software usability. Such use cases are known as essential use cases, because they reduce a task to its essential core, that is, purposes freed of implementation details, technological constraints, and nonessential assumptions. (See my "Essentially Speaking" column, The Unified Process Elaboration Phase, Ambler and Constantine 2000, introducing essential use cases.) Essential use cases make it easier for developers to distinguish the destination from the cow path by which it has been reached in the past. They highlight the working goals of users, separating what users are trying to accomplish from how they might accomplish it. By building use cases based on user intentions, we can often avoid having to implement unnecessary or little-used features.

Requirements worth defining are worth reviewing with clients and users. Not only can reviews help control requirements creep and reduce time pressure, they can also help find errors. Wiegers suggested that agreed-upon requirements be reviewed for testability by developing actual test cases. Defining and walking through test cases early in the process can speed and simplify validation. This practice also highlights problematic or ambiguous requirements that may need to be altered or abandoned. Here, too, modeling with essential use cases has proved to be an advantage. Good test cases fall out of use cases like rain from a thunderhead. Tracking requirements through the entire development process is crucial for effective project management. When test cases are derived directly from the use cases that define requirements, it becomes easier to gauge progress and to recognize and avoid potential feature creep. Tying test cases to use cases also makes it less likely that important capability will be overlooked until it is too late or too expensive to implement it.

Admittedly, requirements are only part of the software development game, but if our polling is any indication, the requirements muddle is a major hole in the middle of the hand of cards that many managers are trying to play. Certainly, none of us on the panel labored under any illusion that we had solved all the problems during the first Management Forum — Live! Nevertheless, in a surprisingly short time, the group offered a panoply of proven practices.

So remember: Define requirements but don't over-define. Educate your clients and users regarding costs and trade-offs. Concentrate on the core. Avoid scope creep by deferring requirements. Build client comfort and confidence through reliable revision and release. Learn what success means to your clients, and distinguish drivers from constraints. Understand what your users want, but meet their needs. Identify what is essential through use cases. Control and validate using test cases derived from use cases.

And never draw to an inside straight.

4

Chapter 4

Best Practices for the Test Workflow

Introduction

The purpose of the Test workflow is to verify and validate the quality and correctness of your system. Following the enhanced lifecycle for the Unified Process, during the Inception phase you can validate a wide range of project artifacts through inspections and reviews. It is possible to test your requirements model — you can do a user interface walkthrough, a use-case model walkthrough, or even use-case scenario testing. It is possible to test your models and project management artifacts by performing peer reviews and inspections of them.

There are several important best practices that you should apply with respect to the Test workflow. First, recognize that if you can build something, you can test it — you can inspect requirements, review your models, and test your source code. In fact, if something isn't worth testing, then it likely isn't worth developing. Second, recognize that silence isn't golden, that your goal is to identify potential defects, not to cover them up. To be successful at testing, you need to have the right attitude — that it is good to test and that it is even better to find defects.

If it isn't worth testing, then it isn't worth creating.

A third best practice is to test often and test early. There are two reasons for this: (1) we make most of our mistakes early in the life of a project, and (2) the cost of fixing defects increases exponentially the later they are found. Technical people are very good at technical things such as design and coding — that is why they are technical people. Unfortunately, technical people are often not as good at non-technical tasks such as gathering requirements and performing analysis — probably another reason why they are technical people. As mentioned earlier, is important to recognize that the cost of fixing these defects rises the later that they are found (McConnell, 1996; Ambler, 2001). This happens because of the nature of software development — work is performed based on work performed previously. For example, the work of the Elaboration phase is performed based on your efforts during the Inception phase. In turn, your Construction phase efforts are based on the work that you did during the Elaboration phase, and so on. If a requirement was misunderstood, all modeling decisions based on that requirement are potentially invalid, all code written based on the models is also in question, and the testing efforts are now verifying the application against the wrong conditions. If errors are detected late in the project lifecycle, they are likely to be very expensive to fix.

The cost of fixing defects rises exponentially the later they are found during the development of an application.

Fourth, acknowledge that the goal of testing is to find defects. Test suites that do not find defects are very likely to be failures — you haven't developed test cases that were able to find problems in your software, and experience shows that there are always defects. Fifth, acknowledge that your software doesn't have to be perfect, it just needs to be good enough. You may find defects, but choose not to address them, or perhaps to address them sometime in the future.

So how can you test the artifacts that you create during the Inception phase? In Figure 4.1, you see a depiction of the Full Lifecycle Object-Oriented Testing (FLOOT) process pattern (Ambler, 1998a; Ambler, 1998b; Ambler, 1999) for testing and quality assurance. Although testing and quality assurance go hand-in-hand, they are two distinct tasks: (1) testing to determine whether or not you built the right thing, and (2) quality assurance (QA) to determine if you built it the right way. It is important to note that FLOOT is merely the tip of the testing iceberg, and that there is a multitude of testing and quality assurance techniques that you can apply during development (Marick 1995; Siegel, 1996; Binder 1999).

4.1 Why Test?

There is a wide variety of reasons why you want to test your work during the Inception phase. Testing early will:

1. Enable you to find and fix defects as early as possible. When you find a defect, you should attempt to fix it as soon as you can so that the defect does not infect other aspects of your work. A misunderstood requirement will infect the models that are based on that requirement, which in turn will affect the source code based on those models. The earlier a defect is addressed, the less expensive it will be to fix.

2. Help reduce the risk of project failure. Testing provides you with an accurate, although fleeting, perspective on quality. You can use this perspective to judge how well your project is doing and to identify potential problem areas so they may be addressed as early as possible.

3. Help reduce overall development costs. By supporting techniques to validate all of your artifacts, the Test workflow enables you to detect defects in your Inception phase deliverables very early in the development of your software.

4. Provide input into the project viability assessment. The purpose of the project viability assessment, a key activity of the Project Management workflow (Chapter 5), is to determine whether or not it makes sense to continue working on your project. An important piece of information to gather in your testing efforts is to determine if your initial understanding of the requirements accurately reflects the needs of your users.

Figure 4.1 Full Lifecycle Object-Oriented Testing (FLOOT).

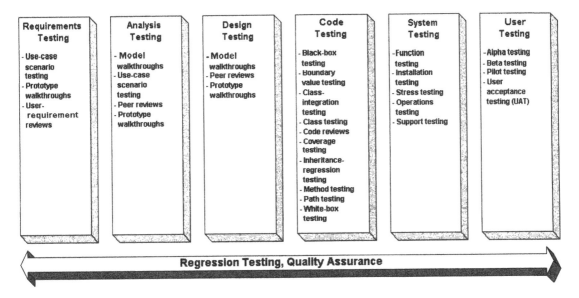

In "A Business Case for QA and Testing" (February 1999, Section 4.4.1), Nicole Bianco argues that testing that begins at the start of a development cycle requires less time over the long haul, but that many organizations don't see things this way. In this article, Bianco provides the following hard numbers regarding testing:

- 44% of the defects occurred during the requirements phase.
- 32% occurred during the design phase.
- Testing tools pay off. Organizations with no or limited testing tools will spend 30%–35% of their development time in testing-related activities. Organizations that have implemented effective testing tools decrease their testing time to 23%–28% of development efforts.

Bianco provides many interesting insights into testing. For example, in one anecdote she describes how a misunderstood requirement was the cause of delivering a system late in one organization — something all of us would prefer to avoid given the opportunity. She shows how the number of software defects could be decreased simply by developing better methods of defining and writing requirements, the topic of Chapter 3. In short, this article provides many valuable insights and best practices that your projects will benefit from.

4.2   Starting Testing Off on the Right Foot

Quality means something different for each project. Some produce systems that are considered high quality because they have low defect levels, they meet a specific customer-requested feature set, or they meet a specific time to market. Most software projects have some balanced combination of all three attributes of customer-perceived value. Johanna Rothman, in "Determining Your Project's Quality Priorities" (February 1999, Section 4.4.3), describes a collection of best practices for creating quality software by defining your project's priorities and constructing a plan to successfully meet that criteria. During the Inception phase, you should strive to define what "quality" means for your project by identifying the defect level, the feature set, and the time to market that your project must meet. You also need to define the release criteria, based on your project's quality definition, so you will know when you're done. Finally, Rothman suggests that you develop a plan that reflects your release criteria. (Planning is an activity of the Project Management workflow, the topic of Chapter 5.)

If you haven't defined what quality means for your project,
how will you know that you've achieved it?

In "Plan Your Testing" (April 1999, Section 4.4.3), Robin Goldsmith describes how to plan your testing efforts so you can find and fix software problems long before your users do. In the article, Goldsmith describes a collection of testing best practices such as:

- dentifying specific tests so you can estimate time and resources to carry out those tests,
- building commitment with project stakeholders to ensure the tests are carried out,
- identifying previously undetected causes of major problems,
- distinguishing and prioritizing the most important tests to ensure that your testing efforts are focused appropriately,
- recognizing that planning tests first can help developers write the code correctly the first time, thereby reducing development time, a fundamental principle of eXtreme Programming (XP) (Beck, 2000),
- ensuring that important tests are run, and run early,
- recognizing that testing is the primary means for reducing risk, and
- recognizing that test planning enables you to apply resources selectively to gain the most "bang for your buck."

Plan to test. Test to your plan.

4.3 Testing Techniques for the Inception Phase

A very simple and straightforward way to validate your artifacts, particularly your requirements model, is to run use-case scenarios (also known as usage scenarios) against them. In "Reduce Development Costs with Use-Case Scenario Testing" (July 1996, Section 4.4.4), Scott Ambler describes a technique called use-case scenario testing (Ambler, 1998a; Ambler, 2001) in which a group of subject matter experts (SMEs), often project stakeholders, are led through the process of validating a business model. Although the article focuses on the validation of a Class Responsibility Collaborator (CRC) model, the technique can easily be generalized to UML class or component models. CRC models are described in detail in the article "CRC Cards for Analysis" (October 1995, Section 2.1.5) by Nancy Wilkinson. Table 4.1 summarizes the advantages and disadvantages of this best practice.

Table 4.1   Advantages and disadvantages of use-case scenario testing.

Advantages	Disadvantages
1. You inexpensively find and fix errors in your models early in the project lifecycle. 2. It provides you with a detailed description of the business logic of the system. 3. It's simple and it works. 4. Your scenarios are a very good start at a use acceptance test plan. 5. It helps to build rapport between project stakeholders.	1. Your SMEs must make the time to do the testing. 2. It's low-tech — developers are often skeptical. 3. Managers often feel that "real" work isn't being accomplished.

An ounce of prevention is worth a pound of cure.

Reviews, a process where people inspect one or more artifacts, is a common best practice that your project team may apply to almost anything that it creates during the development of a system. In "The Seven Deadly Sins of Software Reviews" (March 1998, Section 4.4.5), Karl Wiegers describes seven common problems that can undermine the effectiveness of software technical reviews of any type and several symptoms of each problem. The seven deadly sins are:

1. Participants don't understand the review process.
2. Reviewers critique the producer, not the product.
3. Reviews are not planned.
4. Review meetings drift into problem solving.
5. Reviewers aren't prepared.
6. The wrong people participate.
7. Reviewers focus on style, not substance.

> ***Reviews don't slow projects down, defects slow projects down.***
> ***— Gerald Weinberg***

Luckily, Wiegers also suggests several possible solutions that can either prevent or correct each problem. By laying the foundation for effective software technical reviews and avoiding these common pitfalls, you too can reap the benefits of this valuable quality practice. By being aware of these common problems, and by applying the appropriate solutions, your project team can effectively validate a wide variety of development artifacts, including those developed during the Inception phase.

> ***Inspections without trained moderators will only have a***
> ***moderate success. — Tom Gilb***

4.4   The Articles

4.4.1 "A Business Case for QA and Testing" by Nicole Bianco
4.4.2 "Determining Your Project's Quality Priorities" by Johanna Rothman
4.4.3 "Plan Your Testing" by Robin Goldsmith
4.4.4 "Reduce Development Costs with Use-Case Scenario Testing" by Scott Ambler
4.4.5 "The Seven Deadly Sins of Software Reviews" by Karl Wiegers

4.4.1 "A Business Case for QA and Testing"

by Nicole Bianco

How good is your organization at performing software testing? Consider the following:
- On the first days of a software release, do you sit by the phone waiting for it to ring?
- Do developers arrive at work early during the first week to verify the results of the system's operation so the customer isn't the first one to find defects?
- Do you exert a lot of effort fixing problems after a software release?
- Have you recovered lost or damaged data after the customer has used the software?
- Have you ever told users they will need to repair corrupt data caused by faulty software because there is no software program to fix the data?
- Do you have a lot of people fixing existing software rather than developing new software?

If you answered "yes" to any of these questions, you should evaluate how a formal model of testing could improve your product.

Testing software before releasing it to the user just makes good business sense. However, many software development organizations look at testing as a follow-up to the development mode. With proper testing methods, you can avoid many defects that would otherwise be released into the user's system. An adequate testing life cycle helps you deliver software earlier and with higher quality than if you didn't use a defined model. Delivering higher quality software can help your organization better manage its personnel because there is less work spent repairing failed software.

Cost of Defective Software

To understand the business impact of defective software, you must know its cost. The cost is directly related to the amount of time it takes to find and repair a defect. Here, I will use time as a cost. As an example, I will use a complex system consisting of 750,000 lines of code. The system's organization found that in one year, it spent more than 3,000 person hours repairing the software's problems.

The organization had a sophisticated testing life cycle model, and the numbers represent the effort required to repair defects from all testing phases: from the end of the construction phase through the software's release and operation. Its life cycle model included four phases for testing: unit testing (during the construction phase), integration testing (where the module interfaces are tested), system testing (where the modules are put together and tested), and beta testing (at a customer's site).

The effort required to repair defects equaled the cost of a developer hour, conservatively $75 (including salary, office space, benefits, and so forth), times the 3,000 hours spent fixing problems — a $225,000 expense. With a mature testing life cycle, this organization's results were high-quality software, usually experiencing less than one defect per month. Quality assurance managers estimate that up to half of the defects they found during their development and testing life cycle could have been released into the software if they had used a "common sense" form of testing. This would have increased their defect repair time to about 4,530 hours, or a $340,000 cost.

There was no way to estimate the cost of the customer impact, but it added to the cost of defects — not to mention the customer's frustration with the inability to depend on the software that supports its business functions. In this case, the customer has outsourced more than a few software development functions in the hopes of providing more reliable and stable solutions to business operations.

Table 4.2 shows the one-year history of defect repair for all testing phases in this system development, and defects that were repaired after the software was released, which I'll call "escaped" defects. A total of 2,799 hours were spent removing 335 defects prior to release, an average of 8.4 hours to repair each defect. Any defects that escaped to the customer took an average of 17 hours to repair — that's double the work effort.

Table 4.2   Defects found in the different phases of testing.

Phase	Number of Defects	Repair Hours	Average Time per Defect (hours)
Unit Testing	31	249	8
Integration Testing	160	797	5
System Testing	122	1,313	11
Beta Testing	22	440	20
"Escaped" (after release)	12	206	17

In Table 4.2, you'll also notice that it costs less to remove a defect earlier in the development cycle. If a development hour costs $75, and the average time to repair a defect in the unit testing time is eight hours, the cost of a defect is $600. If you remove the defect in the system testing phase, the cost escalates to $825. Time and dollars are reduced by 27% for each defect found during unit testing instead of later in system testing.

You can decrease the cost of testing (in duration and dollars) with carefully planned training of your testing staff. A process or tool can help you apply the training to assure the greatest impact. For each defect found, ask the developer to track its source. Sources should be defined by phase of development, and never by person. The key phases of development are requirements, definition, design, and construction (coding).

Once you've determined the source of defects, you can train people accordingly. For example, in Table 4.2, 44% of the defects occurred during the requirements phase, and 32% occurred during the design phase. By developing better methods of defining and writing requirements, the number of software defects could be decreased. There are also tools to support these functions (such as requirements management tools or simulation modeling tools) that can help decrease the human resources you use.

Testing that begins at the start of a development cycle requires less time over the long haul, but many organizations don't see things this way. A quantifiable model for a testing life cycle can save you time and effort.

The Case for Shorter Testing Cycles

Determining the time spent to repair a defect was not the mission of the organization depicted in Table 4.2. Instead, its development team knew that the testing cycle added about three weeks to the delivery of any system, and the customers wanted their systems delivered sooner. Releasing the system with more defects was not a reasonable option. It was proposed that if they could insert fewer defects during the design and construction phases, the testing cycle could be shortened. In the case mentioned previously, if no defects were inserted into the software at all (not realistic, but not a bad goal), they would have saved more than 3,000 person hours.

They started by collecting information about each defect they found. They already had a defect tracking method. They logged the following data on each defect: the phase in which the defect was found; the phase in which the defect was inserted; and the amount of time it took to fix the defect.

After a year of collecting the data, the information was analyzed two ways. First, they determined that maintenance — repairing defects and enhancing the existing system's operation after the customer starts using it — caused the defects that took the longest time to repair. During maintenance, seven defects were inserted into the software, each accounting for an average repair effort of 20 hours. Research showed that most of these defects came from fixing other problems. It also showed that many times, fixes were occurring without using the entire test set. Had the test set been run, the subsequent defects would not have occurred.

Second, they determined that most of the defects, 308, originated in the coding (construction) phase. Out of these, 104 were found in testing phases. The construction phase was the largest cause of testing costs and time. Since they averaged 8.4 hours to remove a defect, 873 hours were spent in the testing phases alone. By improving the generation of code, they could save more than 20 person weeks from the development effort and decrease the time to deliver new systems. The organization is establishing standards such as code reviews, training for people who write code, establishing standards for coding, and the use of code generators for some of the construction effort.

Using Tools to Remove Defects

By improving software testing methods, your team can decrease the cost of removing defects. This will improve the degree of customer satisfaction, but will impose additional costs to the development effort. You can use tools designed to support testing activities to decrease the cost of human effort.

There are tools that support regression testing (the testing that exercises the portions of the software that have and have not changed) and tools that will test software interactively while the developer creates the software (interactive debugging tools integrated into the compilers). For real-time systems, there are tools that simulate transactions. Equipment that duplicates the system's operating environment can alleviate the need to schedule testing around the operations environment. Such tools can be costly, and won't normally decrease testing costs, but they can decrease the time spent testing.

For an organization that has no or limited testing tools, my comparison shows that 30% to 35% of their development time will be spent in the testing phases. These phases include the definition of the test cases as well as the time to execute them. Organizations that have implemented tools such as these have decreased their testing time to 23% to 28% of the development life cycle.

As I stated previously, my experience shows that the expense of the tools will offset the cost of human effort. For a department, a manager can expect to spend 18% to 32% of the budget on testing resources, including tools and people. The wide range here is directly related to the number of testing activities occurring. For a small department (10 to 12 developers), the cost would be in a higher range than a larger department (80 to 100 developers). The reason is that the cost of the activity is directly related to the number of uses it receives. So, while a small department may be testing only one new system at a time, a larger one may be testing three or more.

Organizations that release high-quality software — software that performs the functions the user requires — tend to have well-defined testing phases within their software development model. They have created a life cycle model for testing that parallels the development life cycle. They have developers and other resources committed to testing the new code, rather than assigning large numbers of people to maintain existing software with the intent of keeping customer satisfaction high.

But Testing Takes Time

I see this pattern happen over and over again. A software system is written and, to meet a schedule, the testing is abbreviated. Many times, software that is still unstable in the testing phases is released to the user in hopes that delivering anything is better than delivering something late. By delaying the system's implementation and integrating testing, you can save time in repairs and lessen the impact to business operations.

In another organization I worked with, a misunderstood requirement (a type of analysis error) was the cause of delivering a system late. The defect was not discovered until the beta test phase. The beta test phase is often the first time the customers see the results of their requests. If execution of the development goes according to plan, this phase occurs a few weeks prior to the "live" date.

In this case, there was no requirement defining the sequence of reports generated from the system. Since this was a human resources system, the design team assumed the sequence of operation should begin with employee identification number, and designed all the databases accordingly. After seeing the system, the customer decided it was critical to have the reports appear by order of employee last name instead.

The IS organization decided to insert sorts into the reporting structure. After expending 54 person hours attempting this solution, the development team discovered that the processing windows were too small to create the reports by the time the customer needed them. The only approach was to restructure the database to index the file by employee last name instead of the identification number.

The result was more than 800 hours of effort and three weeks of lost weekends and evenings for three people to repair this problem. The resulting system was patched before it was released. Documentation that was kept current throughout development was foregone in attempts to deliver the revised system as quickly as possible. The system was released, but many of the operations had not been adequately tested during the rushed system level modifications. The first defect arrived within three hours of the system going live.

Maintenance was difficult because of last-minute changes to the system flow, programs, and file structures. Repairs took longer than they should have because the documentation was no longer valid. The number of post-release defects was high, adding to the customer's frustration. The customer became quite vocal concerning the IS organization's inability to satisfy a simple system need. This could have been avoided if a proper testing process were incorporated into the development life cycle.

Life Cycle Model for Testing

The common sense approach to developing software includes gathering requirements, defining approaches to satisfy them, selecting an approach, designing the software system, constructing, testing, releasing, and maintaining. As previously discussed, the testing phase often occurs at the end of development. However, to provide high-quality software systems, the testing function should take a parallel path to the development effort. This is accomplished by establishing test cases and scenarios at each development phase. This will provide a test suite that will exercise what the customer requested and provide a more stable implementation.

There can be from two to four or more testing phases, depending on the criticality of the software's functions and the organization's desire to produce error-free software. The organization I discussed uses four phases, but the effort to define the testing scenarios starts with the requirements definition phase. Figure 4.2 shows the model for development.

Once the requirements are well-established, testers (working with users under ideal circumstances) use the written requirements to define a set of tests that will assure the functions requested will work exactly as the requirement is stated. (Had this been done in the case of the wrong sequence report described earlier, it would have saved more than 800 hours and three weeks of delay.) You can use these test cases in the system testing and beta testing phases.

Figure 4.2   A development model for testing.

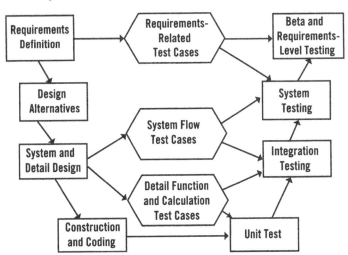

The design alternatives phase, or the approach for implementing the system, will probably not generate test cases unless additional requirements for system operation are included to enhance the system's functions; for example, when the selected design alternative causes a specific platform, development tool, or host operating system to be present. In this case, some testing of the operating system, tool, or platform will be required to function in a certain way.

As the system and detail design phases progress, the testers use their resulting documents again to develop test cases and scenarios to check the proper operation of the system process flows and the detail operation of the individual modules of code as they are designed to function. Many times, test case developers will find errors in the design as they map the detail design to the requirements. Finding defects at this phase can save a lot of effort if you repair them before construction begins. System operation test cases will be used in two testing phases: system testing and integration testing. Detail functional and calculation test cases will be used in integration and unit testing.

Detail functional and calculation test cases will be given to the software developer who is constructing the software to aid in that effort. The testers will accept the software for integration testing only when their test cases for the unit test have been satisfactorily executed. The developer will usually add test cases in the unit test phase, as shown in Figure 4.3. Potential problem areas of code may be developed to ensure the code functions as designed.

Figure 4.3   The test case base for final testing.

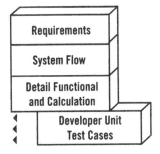

As each testing phase is entered, you add a new set of test cases to exercise the software's functions. All the test cases from the previous phase are added to the ones for the phase being entered, providing a set of regression test cases. By running all previous test cases with the new ones, you assure that the higher levels of system integration are not causing problems in the lower levels of operation. Figure 4.3 shows the entire test case base for final testing.

The Case for Future Tests

No system, once implemented, remains constant. It will need modifications and changes for a number of reasons: changes in business practices, organizational flow, legal requirements, or new sources of input data.

You can reuse the test case set created during the system's development phases to verify that changes made to an existing system do not cause problems anywhere else. You should regression test prior functions for any system change and when a defect is repaired. Likewise, when a defect arises in a system after its release, you should add a new test case to the test case set to ensure it's tested in future modifications. Once the test scenario is available, retesting the changes can exercise the system's functionality, assuring that one enhancement or fix will not impact operation of other functions. You should manage the test case set as a configuration item having the same importance of source code and design documents.

Approach to Quality and Satisfaction

Too many organizations take a casual approach to testing. Organizations that take a more disciplined and methodological approach produce higher-quality products, reduced delivery cycles, and improved levels of customer satisfaction. Also, with practices that encourage and support the creation of high-quality software, developers will take new pride in their work and, with their increased productivity, will help improve the company's business operations.

4.4.2 "Determining Your Project's Quality Priorities"

by Johanna Rothman

Maybe you've worked on an ideal project. You had a project plan with a reasonable schedule, you knew your requirements, and you could do reviews and inspections, build test harnesses, and do exploratory and planned testing.

More often, I've worked on imperfect projects. On these projects, developers and testers feel they don't have the time to do their jobs "right" — reviews and inspections are done incompletely if at all, only exploratory testing is done, the requirements change, and new features are designed and implemented on-the-fly. There is usually an excuse given for these projects — they are under severe time pressure.

But even projects under severe time pressure can be closer to ideal. They can be planned and executed if their quality priorities and release criteria — what success means — are known. In fact, these projects may not need to be under time pressure at all.

Each project is different. Some involve products that are considered high quality because they have low defect levels or a specific customer- or user-requested feature set. Other products are considered high quality if they meet a specific time to market. Most commercial applications and in-house corporate projects have some balanced combination of all three attributes of customer-perceived value. Quality means something different for each project.

Those differences change how you develop and test. But whether you work on in-house corporate projects or commercial projects, you have customers to satisfy.

Consider what success means for each specific project. Customers and users perceive a certain value for each application. That value may have several attributes, and those attributes define quality for your project. You can use quality, the value customers and users perceive, to define success criteria for your projects. Once you've defined the success criteria, you can select a development life cycle for your project. A life cycle will help you plan which activities you do and when to create those quality characteristics.

Defining your project's priorities and constructing a plan to meet success criteria will help your team achieve its desired results — creating quality software.

Define Quality for Your Project

At the onset of a project, you can decide what you need to do by taking the following steps:

- Define quality for this project. What combination of low defect levels, feature set, or time to market does this project need to meet?
- Define release criteria based on your project's quality definition. How do you know when you're done?
- Make a plan that gets you to the release criteria. How do you decide how much of which activities you will complete on this project?

Product quality criteria changes over time. Your customers and previous product releases influence the current project's criteria. I use Geoffrey Moore's *Crossing the Chasm* (Prentice Hall, 1991) high-tech marketing model as a way to consider market forces, and combine that with the software project goals in Robert Grady's *Practical Software Metrics for Project Management and Process Improvement* (HarperCollins, 1992) to come up with a framework for what quality means during a product's lifetime. According to my clients, corporate in-house products follow a similar lifetime.

Table 4.3 combines the market drivers with the project drivers to decide which project quality attributes are important and when. I use this methodology to help drive what quality means to my projects, and then decide which life cycle to use.

Table 4.3 Quality perceptions over product lifetime.

Product Life/ Market Pressure	Introduction (Enthusiasts)	Early Adopters (Visionaries)	Mainstream (Pragmatists)	Late Majority (Conservatives)	Skeptics (Laggards)
Time to Market	High	High	Medium	Low	Low
Feature Set	Low	Medium	Low	Medium*	Medium
Importance of Low Defect Levels	Medium	Low	High	High	High

* Medium market pressure for features does not mean pressure for new features. It means that the promised features must be in the product and must be working.

Enthusiasts want introductory software releases to do a particular thing well, right away. Early adopters need a specific problem fixed, and they want software that does those things reasonably well, right away. Early adopters quickly become power users as they use the software.

Mainstream customers and users want the same content as the early adopters, but they need it to work better than the early adopters because they may not be as skilled in using software or in understanding how the software works. They will wait for the software to be "officially" released, as long as they think it will work for their problem. They will also learn how to use commercial or corporate applications, but they may stay novice users.

Late majority customers will not buy a product or use an application unless you can demonstrate that all of its promised features work reliably. They may remain novice users. Skeptics may buy your product or use your application if you have a good track record with the late majority, and if they perceive your software has features they absolutely require to do their jobs.

Not only does the project have specific customer criteria for quality; there is a bottom-line threshold for defects. Even when time to market is a higher priority than fixing all the defects, most companies prefer not to release damaging products. In the same way, there is an absolute bottom limit to features. A release has to have something new, even if it is just a requirement not to erase disk drives. It might be a minimal requirement, but it exists. Unless management is desperate, it generally decides not to release a damaging or featureless product that could decrease market share or hurt customers. Release-ready software provides some feature or advantage to the company in the marketplace.

Define Release Criteria Based on Quality

Release criteria reveals your project's implicit and explicit critical keys to releasing or not: your project's definition of success or failure. You can define release criteria by deciding what is critical or special to your project. Maybe you have to release by a certain date, need a specific feature or set of features, or you need to track and remove all known defects. Most likely, you have some combination of these concerns. I like to consider the project's time to market, performance, usability, installation, compatibility, defects, and other requirements for the software's release when defining release criteria.

I recently worked with a client, SmartStore Retail Software, that sells into an early adopter marketplace. It needed to have a usable product in a rapid time to market. The developers were used to developing products for the mainstream, where customers will wait to buy the right product. The testers were used to testing products for a late majority market. Neither group thought to test their assumptions against the project requirements. In addition, only some of the feature requirements were specified.

Until the organization developed and agreed upon release criteria, the different groups were frustrated with each other's assumptions about time to market, the feature set, and defect levels. The developers wanted to fix the defects from the previous release before adding new features. Management wanted some new features and some fixes. Testers wanted to focus on performance assessment, not testing new features and fixes. SmartStore could have avoided this non-agreement by defining requirements up front.

Some of SmartStore's final release criteria were:

- Load tests 1, 2, and 3 run at least 5% to 20% faster than they did in the previous release.
- Load tests 4, 5, and 6 run at least 20% to 40% faster than they did in the previous release.

- There are no open, high-priority defects.
- All data exchange functionality is complete (architected, designed, implemented, debugged, and checked in).
- Ships by January 15 (this date meant something specific to this organization and it forced an aggressive project schedule).

For this organization's project, this criteria took the place of specific and traceable requirements. (I don't recommend ignoring requirements and focusing solely on release criteria.)

The criteria focused on getting a reasonable product to market quickly. We negotiated the release criteria with developers, testers, marketing, customer service, and senior management. We all had to agree on the characteristics so we could decide how best to accomplish our goals.

I drafted the initial release criteria. At a project team meeting, I discussed the criteria with the developers and testers to make sure they agreed with me. We discussed each criterion, and whether we thought we could make the criterion by the ship date. The discussion grew heated, so we kept on track by asking these questions for each criterion:

- Must we create this functionality or make this performance?
- What is the effect on our customers if we do not create this functionality or make this performance?

By the end of the project team meeting, we agreed on the criteria and I presented it to the operations committee (senior management, customer service, and marketing management). The committee wanted more functionality, but reluctantly agreed that we were creating a project that provided what they needed.

The developers and testers had to change their actions to create a reasonable product quickly. The developers could no longer just fix existing defects; they had to figure out a way to add more features quickly. Testers couldn't use traditional tools to assess performance, they had to speed up their work to assess performance in addition to testing new features and verifying fixes.

SmartStore had to achieve a specific, short time to market, with a feature set that included performance. Data-damaging defects were not acceptable, but some defects were O.K. SmartStore selected a design-to-schedule life cycle. It completed the architectural design for the next few major releases, and prioritized the development work. The features were divided into three categories: "must," "should," and "walk the dog first." (Each priority was revisited at the planning for the next release.) Some of the "must" work was reflected in the release criteria. "Should" work was not usually mentioned in the release criteria, and the "walk the dog first" work was not mentioned at all.

The testers and developers initially categorized and prioritized the work as they found and entered defects into the defect tracking system. Every day, the engineering management team reviewed and verified the priority of the open defects. We decided on each defect's priority based on how the defect affected the customer or affected our ability to meet the release criteria.

For each of the items in the "must" category, developers did design reviews, code inspection, and unit testing. The testers planned their testing — how much exploratory testing, how much test automation, and when to start regression testing — and had the developers review their test plans.

For the "should" category, the developers tried to do design reviews and code inspections, but more often they ran out of time. The testers tried to plan and develop regression tests, but they always ran out of time. "Walk the dog first" features were never planned or implemented.

SmartStore used a life cycle that helped it reach its release criteria. It knew what it'd get out of development, and it got what it wanted.

Use a Life Cycle that Helps You Reach the Release Criteria

Your project's life cycle defines whether your project will be ideal or imperfect. When the project's life cycle does not match the project's quality priorities, project imperfection occurs.

Different life cycles and techniques have varying effectiveness in terms of the goals of time to market, feature richness, or defect control. Table 4.4 is a comparison of several life cycles and techniques, their strengths, and their product quality priorities. For detailed descriptions of several software development life cycles, see Steve McConnell's *Rapid Development* (Microsoft Press, 1996).

Table 4.4   Life cycles: Their strengths and quality priorities.

Project Life Cycle	Strengths	Priorities
Waterfall: a serial life cycle that in its pure form does not consider feedback. Requires known, agreed-upon, stable requirements, system architecture, and project team.	Requires planning	Feature set Low defect levels Time to market
Spiral: plan, implement, review with customers, plan the next piece.	Risk management	Feature set Low defect levels Time to market
Staged delivery: develop overall architecture, develop in parallel pieces, release when the specified pieces are ready		
Evolutionary delivery: design, review with customer, refine.		
Evolutionary prototyping: small increments of prototype design and implementation, refine, complete, and release.	Deals with rapidly changing or incompletely known requirements.	Time to market Low defect levels Features
Design-to-schedule: develop overall architecture, develop pieces, stop developing at release.		

Project Life Cycle	Strengths	Priorities
Code and fix: not recommended and does not generally meet the desired criteria. Write some code, test, and fix, then loop until project team is too tired to continue.	Deals with and reflects unplanned projects.	Time to market Features Low defect levels

No life cycle is truly appropriate for a first priority of "low defect levels." People tend to think about feature sets with an attribute of extremely high reliability or low levels of defects. Especially in safety-critical or embedded systems, the reliability is really part of the feature set. If the product doesn't work reliably, it just doesn't work.

When you decide on your first priority of quality (time to market, feature set, or low defect levels), you can deliberately choose the most appropriate life cycle to support that definition of quality. Sometimes you might choose a combination of life cycles to meet the mix of quality attributes needed at the time. In essence, you can tailor the life cycle to produce the kind of quality that matches the overall business requirements.

Providing Quality Applications

Not every project needs to perform all of the best software engineering practices equally. (Successful projects do spend enough time on design.) Especially when you have market pressure for time to market or feature set, choose which activities to perform and when. You may choose to test an application using only exploratory testing. You may choose to inspect only certain code. You may choose to only review some of the designs. As long as you know that you will not achieve perfection with these trade-offs, but you will meet your success criteria in your releases, you will provide value to your customers and users with quality applications.

If you spend a little time defining your project's quality priorities, and then choose release criteria to reflect those priorities, you can select a life cycle for your project based on those priorities. When you choose that life cycle, you can have a working environment that makes for an ideal project.

4.4.3 "Plan Your Testing"

by Robin Goldsmith

When a project runs late, testing often gets cut so you can make the scheduled delivery. But what about the inevitable problems that occur once the product is used? Taking time to understand why these problems happen is the first step toward correcting them.

There's a familiar sequence of events behind this all-too-common scenario: The project manager defines programming tasks, estimates the duration and effort needed to perform the tasks, and schedules them on the project plan. Then, he or she schedules a similar amount of time for testing the programs, with testing scheduled just before the scheduled delivery date. Then, reality strikes. Coding takes longer than expected, and testing gets pushed back. You also must squeeze in additional time to test the unplanned coding. The delivery date remains the same, however, and customers are left to find the problems that testing should have found.

Real Causes

Let's analyze why this happens. The development side of the house drives most software projects. As such, the project plan concentrates on development activities. Activities like designing and writing programs are considered important. Other activities, such as testing, often aren't. When the project crunch occurs, it's easy to curtail those activities. Besides, the eternally optimistic developer always assumes the code will be right.

Similarly, the project manager likely applies proven planning techniques to the development activities. At a minimum, the project manager will identify every module that must be developed or modified. He or she will determine each module's required resources and skill levels, and will estimate the effort and duration. Devoting such attention to defining detailed tasks builds confidence in and commitment to them. It would be inconceivable to omit developing a module that's necessary for the project.

Compare this approach with a more liberal method for planning your testing. At best, project managers will use a rule of thumb to "ballpark" the time needed for testing each module. For example, they assume testing a module will take about the same amount of time as coding it. This method provides a degree of reliability, because the project manager presumably applied sound practices to estimate the coding time.

However, by its nature, a rule-of-thumb estimate is given less credence than a seemingly well-thought-out estimate for coding time. In addition, the module-by-module allocation of time for testing tends to apply to unit testing. It can overlook or shortchange other necessary testing such as integration or load tests. The effect is a diminished commitment to testing estimates.

Sometimes project managers simply lump all testing into one testing task. These lump estimates require even less commitment. Moreover, the time allocated for such lumped testing tasks is often just the time left after developers complete "important" programming tasks. Clearly, in such cases, there is not even a pretense of credibility or commitment to the testing tasks.

Real Effects

When testing is cut short to meet a deadline, whatever bugs testing would have found will afflict users instead. It will cost considerably more to fix these defects than if they were found and fixed before the release. Moreover, all tests are not created equal — as project teams that can't distinguish and selectively allocate resources to the most important tests will discover. These teams spend most of their test time on less important tests.

In general, the more important tests — such as integration tests, which confirm interfaces between two or more modules — usually come after you complete unit tests. Tests to demonstrate the system's ability to handle peak loads and system tests are usually performed last. The errors these tests reveal are often the most serious, yet they're the most likely to be crunched.

What's Needed

To deliver quality projects on time and within budget, you need to reduce the number and severity of defects. Otherwise, the defects necessitate extensive unplanned rework, which in turn increases project cost and duration — and the number of errors that persist after the release. The biggest benefits come from methods that ensure requirements and designs are accurate and complete. However, this article focuses on methods to improve testing the delivered code. To be effective, your method should:

- Identify specific tests that need to be performed so that you can reliably estimate time and resources to carry out those tests.
- Build commitment to ensure the tests are carried out.
- Detect previously undetected causes of major problems.
- Distinguish and prioritize the most important tests.
- Ensure that important tests are run, and run early.

A Solution

My consultancy's most effective method to achieve these objectives is a special type of test planning we call "proactive" testing, because it lets testing drive development.

There are three key characteristics of how we "proactively" plan tests: First, planning tests before coding; second, planning tests top-down; and third, planning tests as a means to reduce risks.

Planning tests before coding. If you create test plans after you've written the code, you're testing that the code works the way it was written, not the way it should have been written. Tests planned prior to coding tend to be thorough and more likely to detect errors of omission and misinterpretation. It takes the same amount to write down test plans no matter when you do it. However, writing them first can save you considerable time down the road.

When test plans already exist, you can often carry out tests more efficiently. First, there's no delay. You can run tests as soon as the code is ready. Second, having the test plan lets you run more tests in the same amount of time, because you are using your time to run the tests on the plan instead of interrupting your train of thought to find test data.

Moreover, planning tests first can help developers write the code right the first time, thereby reducing development time. For example, take this simple specification: The operator enters a customer number at a particular location; the program looks up the customer in the database and displays the customer name at a specific location. Could a competent programmer code that wrong? Of course. What happens when you add the following information: Customer Number C123 should be displayed as "Jones, John P." That's a test case, and it helps the developer code the specification correctly.

Rework during development accounts for up to 40% of development time. Difficulty translating design specifications into appropriate code is a major cause of rework. Having the test plans along with the specifications can reduce rework, such as fixing and retesting, significantly.

Top-down planning. Planning tests top-down means starting with the big picture and systematically decomposing it level by level into its components. This approach provides three major advantages over the more common bottom-up method.

First, systematic decomposition reduces the chance that you will overlook any significant component. Since each lower level simply redefines its parent in greater detail, the process of decomposition forces confirmation that the redefinition is complete. In contrast, it is easy to overlook large and significant components with bottom-up planning.

Second, by structuring test design, you can build and manage them more easily and economically. The test structure lets you reuse and redefine the software structure so you can test it with less effort and less rework. It also lets you apply reviews and automated tools where they will be most effective. Third, top-down planning creates the view you need to enable selective allocation of resources. That is, once the overall structure is defined, the test planner can decide which areas to emphasize and which to give less attention. My consultancy uses risk analysis at each successive level to drive the test plan down to greater detail. For more important areas, define more tests. Figure 4.4 illustrates this structure.

Figure 4.4 Top-down test plan structure.

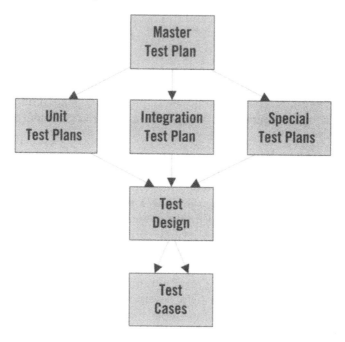

Testing as a means to reduce risks. Testing is the primary means for reducing risk. Therefore, test planning starts with risk analysis to identify and prioritize the risks applicable to the particular test level. At each level, and for each test item, ask the following set of questions to identify risks:

- What must be demonstrated to be confident it works?
- What can go wrong to prevent it from working successfully?
- What must go right for it to work successfully?

Those risks define the components that must be tested at the next level down; and the test plan defines a strategy for accomplishing testing so that higher-priority risks are reduced by testing earlier and more often. The objective is first to do the minimum development necessary to test the high-risk items early. Ensure that key software elements will function properly before building the other elements that depend on them. This way, if the tests reveal problems in the high-risk items, you don't have to throw out or rebuild a lot of software. When you eliminate high risks, you have the time to add and test the lower-risk code more thoroughly.

Master Test Planning

The top level is the Master Test Plan. This is a management document equivalent to, and eventually merged into, the project plan. The Master Test Plan defines the set of Detailed Test Plans for unit tests, integration tests, and special tests which, taken together, will ensure the entire project works. Unit tests deal with the smallest executable pieces of code, usually programs or objects. Ordinarily, the developer unit tests his or her own code. Integration tests test combinations of units and other integrations. Since work from multiple developers is often involved, someone other than the single developer performs integrations tests. Someone other than the developer also typically performs special tests such as load tests, usability tests, and security tests. System tests exercise the complete system end-to-end and are usually performed by someone other than the developer.

While many of you would say you already include such tests in your project plans, I find that "proactive" test planning usually creates a markedly different set of tests from traditional reactive testing.

"Proactive" test planning produces different results. Let's use an accounts payable system as an example. The system design would include modules for maintaining a vendor file and entering orders, but thorough analysis would identify at least the following project-level risks related to these functions:

- Can't find the appropriate vendor when the vendor is already on file.
- Can't add the vendor to the file appropriately.
- Can't modify vendor data.
- Inadequate workflow leads to errors when you enter an invoice for a new vendor or a vendor whose data must be modified.
- Can't enter or modify necessary invoice data accurately.
- Can't identify multiple vendor numbers assigned to one vendor, or has difficulty selecting the proper vendor number.
- Can't transfer an invoice successfully from one vendor number to another.
- Difficulty using the invoice entry or maintenance function, thereby wasting time or increasing errors.
- Enters duplicate invoices.
- Enters invoices for an unauthorized or discontinued vendor.

Traditional project planning would overlook some of these risks, yet they still could occur. You need to test each risk. Also, you want to split out the higher risks so you can code and test them earlier. For example, you might define transferring an invoice as a high risk warranting a separate unit test. Similarly, you might consider some of these risks, such as the ones involving ease of use and procedural workflow, to be higher-priority because of the large impact they could cause. If you were to build both modules and then discover higher-priority risks, it would probably necessitate significant delay and rework.

You can reduce the impact of these higher-priority risks by performing unit tests on them early. You don't need to code the entire modules to carry out these unit tests. Instead, code only the portions of the modules necessary to test the high risks. You may also need stubs or extra scaffolding code to test parts of a module. If the risks don't materialize, you could code the rest of the module. If the risks do show up, you need to redesign the modules before coding; but the total time and effort would be less than if you initially coded the full modules and then recoded them.

Note, by the way, that the risk analysis identifies the need to test manual workflow procedures. Traditional testing probably wouldn't identify such a test because the risk doesn't depend on program coding. However, it is still a major risk to project success and needs to be tested. Risks that you don't identify aren't tested, and they often aren't coded for appropriately, but they do show up.

Lower-Level Test Plans

For each Detailed Test Plan (unit, integration, and special), ask the same three risk analysis questions: What must be demonstrated to be confident the component works? What can go wrong and prevent it from working? What must go right for it to work? Unit test planning will ensure the following risks are tested:

- Can't identify multiple vendor numbers assigned to one vendor or select one from several vendor numbers for that vendor.
- Doesn't identify other vendor numbers for a vendor when you add that vendor.
- Doesn't display all of the vendor's numbers when you enter an invoice.
- Doesn't let you select any of the vendor's numbers when you enter an invoice.
- Doesn't let you correct the selected vendor number when you enter an invoice.
- You need a Test Design Specification for each of these risks. For example, you would need to demonstrate the following to ensure your first Test Design Specification works:
- The other vendor numbers are identified for a vendor when you add that vendor.
- The other vendor number is for the same vendor name
- or address.
- The vendor's other number doesn't share vendor name or business address where one vendor is a subsidiary of the other.
- There are no other vendor numbers for the vendor.
- There is one other vendor number for the vendor.
- The vendor has more than one other vendor number.

For each risk, you must have one or more test cases that, taken together, demonstrate that the risk doesn't occur. Each test case should consist of a specific input and output. You could apply a Test Design Specification to more than one Detailed Test Plan, and a test case to more than one reusable Test Design Specification.

Planned Testing Return on Investment

Top-down test planning based on pre-coding risk analysis can reduce development time and defects. You can test early to ensure that the highest risks do not occur. Moreover, this plan lets you apply resources selectively, to gain the most "bang for your buck."

4.4.4 "Reduce Development Costs with Use-Case Scenario Testing"

by Scott Ambler

Acting out use-case scenarios along with the users of your system will eliminate requirements errors.

We all make mistakes, which is why we need to test the systems we develop. There are two costs to testing: the cost of finding errors and the cost of fixing them. While it is important to reduce the cost of finding errors, it is far more important to reduce the cost of fixing them. When mistakes are finally found during the traditional testing phase, it is often too late or too expensive to fix them, which is why this task is often left until the next release. This article describes a tried and proven technique called use-case scenario testing that can reduce the cost of fixing errors by as much as five orders of magnitude.

Use-case scenario testing is typically performed during the Inception or Elaboration phases of a project by a group of business domain experts (BDEs) (also known as subject matter experts (SMEs) or simply as project stakeholders) — people who are normally users. The conceptual model, which is typically in the form of a CRC (Class, Responsibility, Collaborator) Model or a UML Class Model, is put to the test by the BDEs. By doing this, the BDEs ensure that their requirements definition is accurate and complete, preventing potentially catastrophic errors later in the project. The bottom line is that use-case scenario testing helps to ensure that you have a strong foundation upon which to build your application.

What is a Use-Case Scenario?

A use-case scenario is a description of a potential business situation that users of a system may face. A scenario can be a specific instance of a use case, a portion of a use case, or even a path through several use cases (Ambler, 2001). Acting out these scenarios helps the development team discover if they are missing or misunderstanding user requirements.

For example, the following would be considered use-case scenarios for a university information system:

- A student wishes to enroll in a course but doesn't have the prerequisites for it.
- A professor requests a course list for every course that she teaches.
- A student wants to drop a course the day before the drop date.
- A student wants to drop a course after the official drop date.
- A student requests a printed copy of his transcript so that he can include copies of it with his resume.

As you can see from these examples, use-case scenarios include situations that the system should and shouldn't be able to handle. For example, we have two different "drop course" scenarios: one before the drop date and one after. The reason why we create use-case scenarios that the system isn't expected to handle is simple: system analysis is the task of defining what an application should and shouldn't be able to do. We need use-case scenarios that are potentially out of the scope of the system to help us define its boundaries.

Radical Ideas

Two key concepts lie behind use-case scenario testing. Many experienced developers consider both of them radical — even threatening. The first concept is that users should perform system analysis, not developers. The reasoning is simple — users are the business domain experts. It makes sense to have the people who truly understand the business — the users — perform the analysis of the system.

Although developers have traditionally performed system analysis, experience shows us that this isn't an effective way to operate. The systems we create often don't meet the needs of our users. Although we are experts at technology, we aren't necessarily experts at the business we're creating an application for.

Many developers unfortunately don't see it this way. A simple test to determine if you truly understand the target business is whether or not you can do the jobs of all your users just as well as they can. If you can't, or even if you think you could if you received the right training, then you don't know the business. Class, Responsibility, Collaborator (CRC) modeling enables users to perform an analysis in a few minutes. This technique is explained later in the article.

The second concept behind use-case scenario testing is that we should test while it is still very inexpensive to fix errors — during the early phases of a project. Because the snowballing cost of finding and fixing errors makes it clear that we need to deal with errors sooner in the development lifecycle, this makes a lot of economic sense.

The Roles

Use-case scenario testing is performed by several people who each take on one of four roles:

Business Domain Experts (BDEs):   These are the real users of the system. They do the work day in and day out and have the business domain knowledge. Typically, four or five BDEs are involved in use-case scenario testing. If you have less than four, you'll find you won't get a wide enough range of knowledge and experience. Once you get beyond five or six people, the BDEs begin to get in each other's way. While you must have BDEs who are real users of the system and who know the business, relative "outsiders" who have a vision of the future are also valuable BDEs. These people could be business analysts, system architects, or simply users from a different part of the company.

Facilitator: This person runs the session. The facilitator communicates what the conceptual model (which for the sake of this article I will assume to be a CRC model) and use-case techniques are all about, makes sure that these techniques are done properly, asks pertinent questions during modeling, recognizes when prototyping needs to be done, and leads the prototyping effort. Facilitators need good communication skills and are often either the project leader or a trained meeting facilitator.

Scribes: You should have one or two scribes in the room. Scribes take down the detailed business logic that isn't captured on the CRC cards. Scribes do not actively participate in the session, although they may ask questions to confirm the business rules or processing logic they are recording.

Observers: For training purposes, you may wish to have one or more people sit in on the modeling session as observers. These people sit at the back of the room and do not participate in the session.

Once you have assembled your group, you can begin use-case scenario testing. The testing process involves three steps:

1. Create an initial conceptual model using CRC cards.
2. Create the use-case scenarios.
3. Act out the scenarios.

Creating a CRC Model

CRC modeling is an object-oriented user requirements definition and/or business modeling technique, a primary technique of the eXtreme Programming (XP) methodology (Beck, 2000), in which a facilitator-led group of business domain experts works to define the user requirements for an application.

The technique was originally proposed as an object-oriented training method by Kent Beck and Ward Cunningham in "A Laboratory for Teaching Object-Oriented Thinking" (OOPSLA 1989 Conference Proceedings). CRC modeling quickly caught on in the object-oriented community as a serious modeling technique. The process is mentioned in the books *Design of Object-Oriented Software*, by Wirfs-Brock et al. (Prentice Hall, 1990) and *Object-Oriented Software Engineering, a Use-Case Driven Approach*, by Ivar Jacobson (Addison-Wesley, 1992).

In 1992, Ivar Jacobson, et al. added use cases to the CRC modeling process. Now, it is clear that use-cases can be used at the end of a CRC modeling session to verify that the user requirements definition is correct. Hence, the name "use-case scenario testing." In short, CRC modeling is the precursor of use-case scenario testing.

A CRC model is really a collection of standard index cards that represent classes. The cards are divided into three sections. Along the top of the card you write the name of the class. In the body of the card you list the class responsibilities on the left and the collaborators on the right.

Before you begin CRC modeling, the facilitator must explain two things: what a CRC model is and how to perform CRC modeling. This will take between 10 and 15 minutes and often involves working through an example or two. It is also common for facilitators to distribute a short description of what CRC modeling is all about a few days before the modeling session. This allows the business domain experts to get up to speed on the technique beforehand.

After the explanation, the facilitator leads the BDEs through a brainstorming session in which use-cases are generated. The topic of discussion is naturally the system being analyzed, with a concentration on what the system should do and why. Brainstorming helps the participants get their creative juices flowing and become focused on the task at hand. Once the brainstorming is complete, it is time to create the CRC model. Creating the model is a five-step iterative process.

Step 1: Find the classes.   A class is a person, place, thing, event, or concept relevant to the system at hand. For example, in a university information system there would be student, professor, and course classes.

Finding classes is fairly straightforward. You should look for anything that interacts with or is part of the system. For example, in a university information system, we would have cards for students, professors, and courses. You should also ask yourself where the money comes from (usually customers), how it is earned (through the sale of products or services), and what it is spent on. By following the money, you can identify many core classes for the system (the customers, the products and services being sold, and the components that make up the product and service being sold).

Reports and screen are modeled as classes. For example, one of the reports generated by our system would be a list of all students taking a course. Therefore, we'd have a card called "Class List." Figure 4.5 shows some of the CRC cards that would make up a university information system.

Figure 4.5   CRC cards for a university information system.

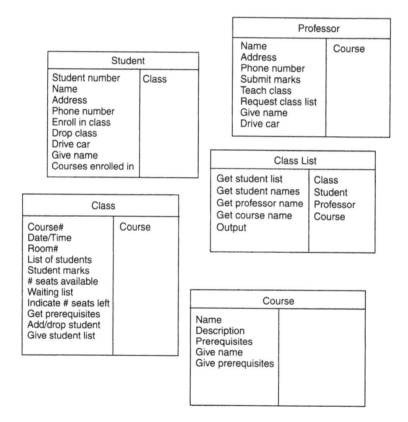

Step 2: Find responsibilities.   A responsibility is anything that a class knows or does. For example, students have names, addresses, and phone numbers. These are the things that a student knows. Students also enroll in courses, drop courses, and request transcripts. These are the things that a student does. The things a class knows and does constitute its responsibilities. A class is able to change the values of the things it knows, but it is not able to change the values of what other classes know. In other words, classes update their own attributes and nobody else's.

Finding responsibilities is just as simple as finding classes. Ask yourself what information you need to store about the class and what the class does. For example, we need to store the names, addresses, and phone numbers of our students. We'll have "Name," "Address," and

"Phone number" responsibilities listed on our "Student" card. Further, because students enroll in and drop out of courses, we'd have "Enroll in course" and "Drop course" responsibilities.

Step 3: Define collaborators.   Collaboration occurs whenever a class has a responsibility to fulfill but doesn't have enough information to fulfill it. For example, students sign up for courses. To do this, a student needs to know if a spot is available. If so, he or she needs to be added to the course. However, students only have information about themselves (their name, address, and so on) and not about the courses. The student needs to collaborate (work with) the card labeled "Class" to sign up for the course.

Collaboration will take one of two forms: A request for information (in this case, the card "Student" requests an indication from the card "Class" whether or not there is a space available) or a request to do something (in this case, "Student" will request to be added to the "Class" if a seat is available). Think of it like this: Collaborators are the other classes that this class is related to or associated with.

You need to define collaborations whenever you find a responsibility that a class can't handle by itself. For example, the class "Student" doesn't have enough information (it doesn't know how many seats are left) to enroll in a course. As a result, it needs to collaborate with Course" to fulfill its "Enroll in course" responsibility. It must first determine how many seats are left in the course and, if there are any left, request that the student be added to the course. The result of this collaboration is that "Class" is added as a collaborator of "Student," and the responsibilities "Indicate number of seats left" and "Add student to course" are added to "Class."

Step 4: Create and arrange the CRC cards.   As the CRC cards are created, the BDEs gather around a large desk and place the cards on the desktop so everyone can see them. Two cards that collaborate with one another should be placed close together on the desk, whereas two cards that do not collaborate should be placed far apart. While this is a small thing to do, it has a significant impact on the CRC modeling process because it helps the BDEs visually grasp the relationships between the various classes in the system.

Step 5: Prototyping.   Prototyping is performed whenever a screen or report class is identified. As soon as a BDE identifies a new screen or report, the facilitator should step in and lead the BDEs through a quick prototyping session, where the screen or report is sketched on either a whiteboard or flip chart. A flip chart is preferable, as it gives you a permanent sketch you can tape to the wall.

You prototype the screens and reports to get a better feel for what the system interface will be like. Further, once you have a prototype, it's easy to identify responsibilities — the data shown or edited on your screens and reports has to come from somewhere, doesn't it?

Pros and Cons of CRC Modeling

The main advantage of CRC modeling is that users perform the analysis of a system. This makes a lot of sense because the users know the business the best. CRC modeling is simple, inexpensive, and portable, leading directly into class modeling, a key technique of the Unified Process's Analysis and Design workflow. The main disadvantage of CRC modeling is that developers are threatened by it. It's hard for experienced systems professionals to believe that something this simple is so effective.

Once your BDEs have created an initial CRC model, it's time to start testing their analysis efforts to verify that the model works. In other words, it is time to start formulating use-case scenarios.

How to Create Use-Case Scenarios

Creating use-case scenarios begins with the facilitator, who should first explain to the BDEs what use-case scenarios are and provide a few examples. As a group, the BDEs should do some brainstorming to identify potential use-case scenarios. The facilitator should lead the brainstorming session, while the scribe writes down the scenarios on flip-chart paper, taping the paper to the walls of the room. Once the brainstorming session is finished, the group transcribes the use-case scenarios onto index cards. Because you can usually describe use-case scenarios with a sentence or two, I have found that you can write each scenario on its own index card. Write the description of the scenario (such as "a student who has the prerequisites wants to enroll in a seminar") along the top of the card. Write the various actions to take (such as "if there is an available seat, enroll them") in the body of the card.

You can break down the use-case scenarios in one of two ways, as shown in Figure 4.6. You can write a generic use-case scenario and describe each alternative action. For example, when a student enrolls in a course, students are automatically allowed to enroll if they meet the prerequisites. If they do not, they must fill out a form explaining why they want to take the class. This form is sent to the professor who teaches the class, and the professor makes the decision whether or not to allow the student to take the course. You would write one use-case scenario for a student who wants to enroll in a course, using the body of the card to describe both alternatives.

Or, you can write specific use-case scenarios for each alternative action relating to the scenario. For the previous example, you would write two use-case scenarios, one for a student with the prerequisites and one for a student without the prerequisites.

As mentioned previously, you should create scenarios that the system should and shouldn't be able to handle so you define what is in and what is out of scope for the system. Further, if your users have told you about a business rule, create a scenario that tests it. For example, if students can take a maximum of five courses, create a scenario for someone trying to enroll in six. By doing this, you'll bring into question the validity of existing business rules that may or may not make sense anymore.

Acting Out Use-Case Scenarios

Once you have defined the use-case scenarios, the BDEs must act them out to determine what changes, if any, need to be made to their CRC model. Led by the facilitator, the BDEs act out the scenarios one at a time, taking on the roles of the their cards and describing the business or processing logic of the responsibilities supporting each use-case scenario. The scribes write this information down as the BDEs describe it. These verbal descriptions enable the BDEs to gain a better understanding of the business rules and logic of the system and to find missing or misunderstood responsibilities and classes.

Figure 4.6 Two use-case scenario formats.

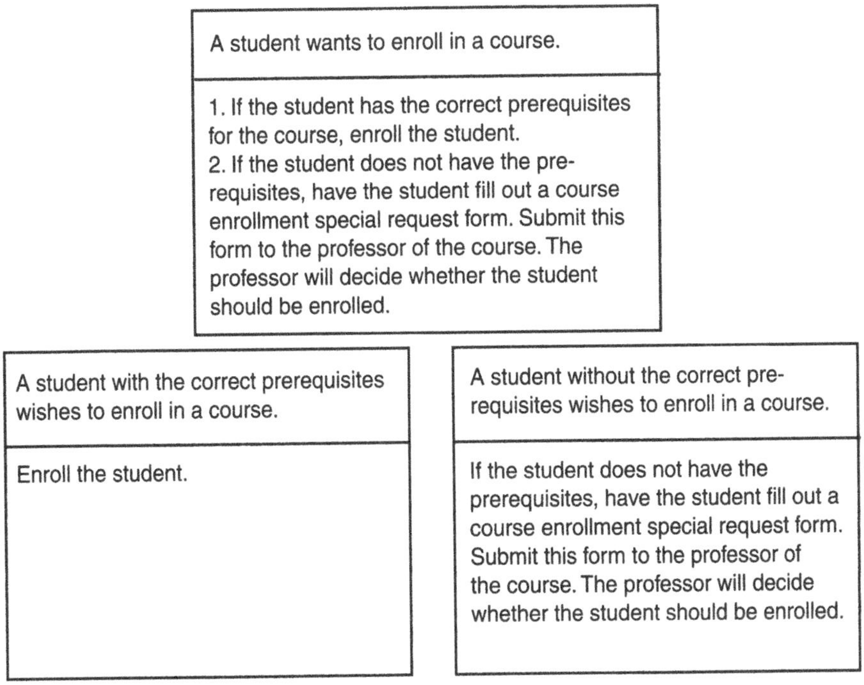

A student wants to enroll in a course.
1. If the student has the correct prerequisites for the course, enroll the student. 2. If the student does not have the prerequisites, have the student fill out a course enrollment special request form. Submit this form to the professor of the course. The professor will decide whether the student should be enrolled.

A student with the correct prerequisites wishes to enroll in a course.	A student without the correct prerequisites wishes to enroll in a course.
Enroll the student.	If the student does not have the prerequisites, have the student fill out a course enrollment special request form. Submit this form to the professor of the course. The professor will decide whether the student should be enrolled.

To act out a scenario, the CRC cards are distributed evenly among the BDEs. Each BDE should have roughly the same amount of processing in their hands. Some BDEs will have one or two "busy" cards, while others may have numerous not-so-busy cards. The main goal here is to spread the functionality of the system evenly among the BDEs. It is very important that you do not give two cards that collaborate to the same person. When you see how use-case scenarios are acted out, you'll understand why.

The person with the card in question holds a soft, spongy ball, which indicates which card is currently "processing." Whenever a card has to collaborate with another one, the user holding that card throws the ball to the holder of the collaborating card.

The ball helps the group keep track of who is currently describing the business logic and also helps make the entire process a little more interesting.

Figure 4.7 shows the logical sequence of acting out use-case scenarios:

1. Call out a new scenario. The facilitator calls out the description of the scenario and the actions to be taken. The group must decide if this scenario is reasonable (remember, the system can't handle some scenarios), and, if it is, which card is initially responsible for handling it. For example, with the "Enroll a student with the prerequisites in a seminar" scenario, the card that would be responsible for initiating this scenario would be "Student." The facilitator starts out with the ball and throws it to the person holding that card. When the scenario is completely acted out, the ball will be thrown back to the facilitator.

2. Determine which card should handle the responsibility. When a scenario has been described, or when the need for collaboration has been identified, the group should decide which CRC card should handle the responsibility. Often, a card identifying such responsibility will already exist. If this isn't the case, update the cards. Once the update is complete, whoever has the ball should throw it to the person holding the card listing that responsibility. In our example, the ball would be thrown first to the person holding the "Student" CRC card.

Figure 4.7   Acting out use-case scenarios.

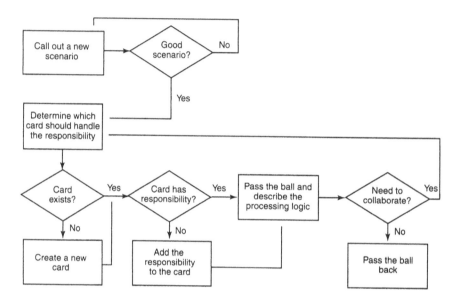

3. Update the cards when necessary.   When you discover new responsibilities, add them to an existing card or create new cards for them. For example, on our CRC model, we might discover that we are missing a "Check for prerequisites" responsibility on the "Enrollment Screen" card. If a responsibility needs to be added to an existing card, find the card that would logically include that responsibility and update it. If a new card needs to be created, the facilitator should hand a blank CRC card to one of the BDEs and ask him or her to fill it out. The BDEs should also be responsible for updating the cards in their possession. You may also need to update your prototype drawings — if an interface or report class changes, the prototype may need to change as well.

4. Describe the processing logic.   When the ball is thrown to someone, he or she should describe the business/processing logic for the responsibility step-by-step. For example, to describe the business logic of verifying the prerequisites, we might say that we determine what the prerequisites are, determine what seminars the student has taken in the past, and then compare the two lists to see that the prerequisites are all there.

Think of it like this: The BDEs effectively describe pseudocode (high-level program code) for the responsibility. This is often the most difficult part of use-case scenario testing. If some BDEs don't think logically, the facilitator must help them through the logic.

You'll find that after running through the first few scenarios, the BDEs will quickly get the hang of describing processing logic.

The scribe should write down these descriptions (remember, the job of the scribe is to record the business logic and rules for the system, which is exactly what the BDE is currently describing.) The facilitator ensures that the processing logic is being described thoroughly, that the BDEs update the cards, and that they throw the ball whenever they need to collaborate. It's often tempting to use a tape or video recorder to take down the business logic. The time lost for transcription (between two and six hours per hour of tape) plus the fact that people don't like to be recorded should make you think twice about this alternative.

4. Collaborate when necessary. To verify the prerequisites, the "Enrollment Screen" card needs to collaborate with "Course" to find out what the prerequisites are and then collaborate with "Student" to find out what courses have already been taken. As the BDE describes the business logic of the responsibility, he or she will often describe a step that needs to collaborate with another card. That's O.K. If this is the case, go back and determine which card should handle the responsibility.

5. Pass the ball back when done. Eventually, the BDE will finish describing the responsibility. When this happens, he or she should throw the ball back to the person who originally threw it. This will be either another BDE or the facilitator.

6. Save the scenarios. Don't throw the scenarios away once you are finished acting them out. The stack of scenario cards is a good place to begin your user acceptance test plan.

The Pros and Cons

Table 4.5 summarizes the pros and cons of use-case scenario testing. The advantages are that it produces close to a bulletproof CRC model, a detailed description of the business logic for the system as documented by the scribes, and a set of use-case scenario cards.

Table 4.5 Pros and cons of use-case scenario testing.

Pros	Cons
• You inexpensively find and fix analysis errors.	• Your BDEs (users) must make the time to do the testing.
• It provides you with a detailed description of the business logic of the system.	• It's low tech — developers are often skeptical.
• It's simple and it works.	• Managers often feel that "real" work isn't being accomplished.
• Your scenarios are a very good start at a use acceptance test plan.	

The CRC model provides a detailed overview of the system, which when supported by the detailed business logic recorded by the scribes provides a perfect start to a class model. Further, we know that our CRC model accurately reflects the business because we've already tested it. The collection of use-case scenarios provides an excellent start at the formulation of a user acceptance test plan. In other words, although use-case scenario testing is a rather simple and straightforward technique, it produces some very significant results.

Unfortunately, use-case scenario testing has several minor disadvantages. First, the BDEs must take the time to do the testing, which can easily take several days. Second, developers are often skeptical about use-case scenario testing due to its simplicity. Finally, because you have a group of users throwing a ball around and filling in index cards, managers often feel that "real" work isn't being done.

Testing Saves Time and Money

Analysis errors are the most common and the most destructive type of mistakes. Use-case scenario testing is a simple and inexpensive method to find and fix these errors during the analysis process, which is the least expensive time in the system development lifecycle to do so.

Use-case scenario testing reduces the cost of fixing errors by several orders of magnitude. These cost savings result because no errors are coded into the system in the first place. Use-case scenario testing proves that there is some truth in the old adage, "An ounce of prevention is worth a pound of cure."

4.4.5 "The Seven Deadly Sins of Software Reviews"

by Karl Wiegers

A quarter-century ago, Michael Fagan of IBM developed the software inspection technique, a method for finding defects through manual examination of software work products (anything produced on a project, such as code, documentation, and so forth) by a group of the developer's peers. Many organizations have achieved impressive results from inspections, including Raytheon, Motorola, Bell Northern Research, Hewlett Packard, and Bull HN Information Systems. However, other organizations have had difficulty getting any kind of software review process going. Considering that effective technical reviews are one of the most powerful software quality practices available, all software groups should become skilled in their application.

In this article, I will describe seven common problems that can undermine the effectiveness of software technical reviews of any type (inspections being a specific type of formal review) and several symptoms of each problem. I'll also suggest several possible solutions that can prevent or correct each problem. By laying the foundation for effective software technical reviews and avoiding these common pitfalls, you too can reap the benefits of this valuable quality practice.

Sin 1: Participants Don't Understand the Review Process

Symptoms. Many software engineers don't instinctively know how to conduct and contribute to software reviews. Review participants may have different understandings of their roles and responsibilities, and of the activities performed during a review. Team members may not know which of their work products should be reviewed, when to review them, who should participate, and what review approach is most appropriate in each situation.

Team members may not understand the various types of reviews that can be performed. The terms "review," "inspection," and "walkthrough" are often used interchangeably, although they are not the same beast. A lack of common understanding about review practices can lead to inconsistencies in review objectives, review team size and composition, forms used, record keeping, and meeting approaches. Too much material may be scheduled for a single review because participants are not aware of realistic review rates. It may not be clear who is running a review meeting, and meetings may lose their focus, drifting from finding defects to solving problems or challenging the developer's programming style. The consequences of this confusion are typically missed defects, frustration, and an unwillingness to participate in future reviews.

Solutions. Training is the best way to ensure that your team members share a common understanding of the review process. For most teams, four to eight hours of training should be sufficient, though you may wish to obtain additional specialized training for those who will act as inspection moderators. Training can be an excellent team-building activity, as all members of the group hear the same story on some technical topic and begin with a shared understanding and vocabulary.

Your group should also adopt some written procedures for how to conduct reviews. Procedures help review participants understand their roles and activities, so they can consistently practice effective and efficient reviews.

Your peer review process should include procedures for both formal and informal reviews. Not all work products require formal inspection (though inspection is indisputably the most effective review method), so a palette of procedural options will let team members choose the most appropriate tool for each situation. Adopt standard forms for recording issues found during review meetings, and for recording summaries of the formally conducted reviews. Good resources for guidance on review procedures and forms are *Software Inspection Process* (McGraw-Hill, 1994) by Robert Ebenau and Susan Strauss and *Handbook of Walkthroughs, Inspections, and Technical Reviews, 3rd Edition* (Dorset House, 1990) by Daniel Freedman and Gerald Weinberg.

Sin 2: Reviewers Critique the Producer, Not the Product

Symptoms. Initial attempts to hold reviews sometimes result in personal assaults on the skills and style of the work product author. A confrontational style of raising issues exacerbates the problem. Not surprisingly, this makes the developer feel beaten down, defensive, and resistant to legitimate suggestions that are raised or defects that are found. When developers feel personally attacked by other review participants, they will be reluctant to submit their future products for review. They may also look forward to reviewing the work of their antagonists as an opportunity for revenge.

Solutions. When helping your team begin reviews, emphasize that the correct battle lines pit the developer and his or her peers against the defects in the work product. A review is not an opportunity for reviewers to show how much smarter they are than the developer, but rather a way to use the collective wisdom, insights, and experience of a group of peers to improve the quality of the group's products. Try directing your comments and criticisms to the product itself, rather than pointing out places the author made an error. Practice stating your comments in the first person: "I don't see where these variables were initialized," not "You forgot to initialize these variables."

In an inspection, the roles of the participants are well-defined. One person — not the author — is the moderator and leads the inspection meeting. In the review courses I teach, students often ask why the author does not lead the inspection meeting. One reason is to defuse the confrontational nature of describing a defect directly to the person who is responsible for the defective work product. I have found the best results come when both reviewers and author check their egos at the door and focus on improving the quality of a work product.

Sin 3: Reviews Are Not Planned

Symptoms.   On many projects, reviews do not appear in the project's work breakdown structure or schedule. If they do appear in the project plan, they may be treated as milestones rather than tasks. Because milestones take zero time by definition, the non-zero time that reviews actually consume may make the project appear to slip off schedule because of them. Another consequence of not planning reviews is that potential review participants often haven't set time aside to participate.

Solutions.   A major contributor to schedule overruns is inadequate planning of the tasks that must be performed. Not including these tasks in the schedule doesn't mean you won't perform them; it simply means that when you do perform them, the project may wind up taking longer than you expect. The benefits of well-executed software technical reviews are so great that project plans should explicitly show that key work products will be examined at planned checkpoints. Slippages more often happen because of the defects found, not because of the review time.

When planning a review, estimate the time required for individual preparation by the reviewers, the review meeting (if one is held), and likely rework. (The unceasing optimism of software developers often leads us to forget about the rework that follows most quality assurance activities.) The only way to create realistic estimates of the time needed is to keep records from your reviews of different work products. For example, you may find that your last 20 code reviews required an average of five staff-hours of individual preparation time, eight staff-hours of meeting time, and seven staff-hours of rework. Without collecting such data, your estimates will forever remain guesses, and you will have no reason to believe you can realistically estimate the review effort on your future projects.

Sin 4: Review Meetings Drift into Problem Solving

Symptoms.   Software developers are creative problem solvers by nature. We enjoy nothing more than sinking our cerebrums into sticky technical challenges, exploring elegant solutions to thorny problems. Unfortunately, this is not the behavior we want during a technical review. Reviews should focus on finding defects, but too often an interesting bug triggers a spirited discussion about how it ought to be fixed.

When a review meeting segues into a problem solving session, the progress of examining the product grinds to a halt. Participants who aren't equally fascinated by the problem at hand may become bored and tune out. Debates ensue as to whether a proposed bug really is a problem, or whether an objection to the developer's coding style indicates brain damage on the part of the reviewer. Then, when the reviewers realize the meeting time is almost up, they hastily regroup, quickly flip through the remaining pages, and declare the review a success. In reality, the material that is glossed over likely contains some major problems that will come back to haunt the development team in the future.

Solutions.   The kind of reviews I'm discussing in this article have one primary purpose: to find defects in a software work product. Solving problems is usually a distraction that siphons valuable time away from the focus on error detection. One reason inspections are more effective than less formal reviews is that they have a moderator who controls the meeting, including detecting when problem solving is taking place and bringing the discussion back on track. Certain types of reviews, such as walkthroughs, may be intended for brainstorming, exploring design alternatives, and solving problems. This is fine, but don't confuse a walkthrough with a defect-focused review such as an inspection.

My rule of thumb is that if you can solve a problem with no more than one minute of discussion, go for it. You have the right people in the room and they're focused on the issue. But if it looks like the discussion will take longer, the moderator should remind the recorder to note the item and ask the author to pursue solutions after the meeting.

Rarely, you may encounter a show-stopper defect, one that puts the whole premise of the product being reviewed into question. Until that issue is resolved, there may be no point in completing the review. In such a case, you may choose to switch the meeting into problem-solving mode; but if you do that, don't pretend you completed the review as originally intended.

Sin 5: Reviewers Aren't Prepared

Symptoms.   You arrive at work at 7:45 a.m. to find a stack of paper on your chair with a note attached: "We're reviewing this code at 8:00 a.m. in conference room B." There's no way you can properly examine the work product and associated materials in 15 minutes. If attendees at a review meeting are seeing the product for the first time, they may not understand the intent of the product or its assumptions, background, and context, let alone be able to spot subtle errors. Other symptoms of inadequate preparation are that the work product copies brought to the meeting aren't marked up with questions and comments, and some reviewers don't actively contribute to the discussion.

Solutions.   Since about 75% of the defects found during inspections are located during individual preparation, the review's effectiveness is badly hampered by inadequate preparation prior to the meeting. This is why the moderator in an inspection begins the meeting by collecting the preparation times from all participants. If the moderator judges the preparation time to be inadequate (say, less than half the planned meeting time), he or she should reschedule the meeting. Make sure the reviewers receive the materials to be reviewed at least two or three days prior to the scheduled review meeting.

When reviews come along, most people don't want to interrupt their own pressing work to carefully study someone else's product. Try to internalize the fact that the time you spend reviewing a coworker's product will be repaid by the help you'll get when your own work comes up for review. Use the average collected preparation times to help reviewers plan how much time to allocate to this important stage of the review process.

Sin 6: The Wrong People Participate

Symptoms.   If the participants in a review do not have appropriate skills and knowledge to find defects, their review contributions are minimal. Participants who attend the meeting only to learn may benefit, but they aren't likely to improve the quality of the product. Management participation in reviews may lead to poor results. If the team feels the manager is counting the bugs found to hold against the developer at performance appraisal time, they may hesitate to raise issues during the discussion that might make their colleague look bad.

Large review teams can also be counterproductive. I once participated in a review (ironically, of a draft peer review process) that involved 14 reviewers. A committee of 14 can't agree to leave a burning room, let alone agree on whether something is a defect and how a sentence should properly be phrased. Large review teams can generate multiple side conversations that do not contribute to the review objectives and slow the pace of progress.

Solutions. Review teams having three to seven participants are generally the most effective. The reviewers should include the author of the work product, the author of any predecessor or specification document for the work product, and anyone who will be a user of the product. For example, a design review should include (at least) the designer, the author of the requirements specification, the implementor, and whoever is responsible for integration testing. On small projects, one person may play all these roles, so ask some of your peers to represent the other perspectives. It's OK to include some participants who are there primarily to learn (an important side benefit of software reviews), but focus on people who will spot bugs.

I'm not dogmatic on the issue of management participation. As a group leader, I also wrote software. I needed to have it reviewed (thereby setting an example for the rest of the team), and I was able to contribute usefully to reviews of other team members' products. This is very much a cultural issue dependent upon the mutual respect and attitudes of the team members. A good rule of thumb is that only a first-line manager is permitted in a review, and then only if it is acceptable to the author. Managers should never join in the review "just to see what's going on."

Sin 7: Reviewers Focus on Style, Not Substance

Symptoms. Whenever I see a defect list containing mostly style issues, I'm nervous that substantive errors have been overlooked. When review meetings turn into debates on style and the participants get heated up over indentation, brace positioning, variable scoping, and comments, they aren't spending energy on finding logic errors and missing functionality.

Solutions. Style can be a defect if excessive complexity, obscure variable names, and coding tricks make it hard to understand and maintain the code. This is obviously a value judgment: an expert programmer can understand complex and terse programs more readily than someone with less experience. Control the style distraction by adopting coding standards and standard templates for project documents. These will help make the evaluation of style conformance more objective. Use code reformatting tools to enforce the standards so people can program the way they like, then convert the result to the established group conventions.

Be flexible and realistic about style. It's fine to share your ideas and experiences on programming style during a review, but don't get hung up on minor issues, and don't impose your will on others. Programming styles haven't come to Earth from a font of universal truth, so respect individual differences when it is a matter of preference rather than clarity.

A Little Help from Your Friends

Any software engineering practice can be performed badly, thereby failing to yield the desired benefits. Technical reviews are no different. Many organizations have enjoyed excellent results from their inspection programs, with returns on investment of up to tenfold. Others, though, perceive reviews to be frustrating wastes of time, usually because of the seven deadly sins I mentioned here, not because reviews are inherently a waste of time. By staying alert to these common problems and applying the solutions I've suggested, you can help ensure your technical review activities are a stellar contributor to your software quality program.

Chapter 5

Best Practices for the Project Management Workflow

Introduction

The purpose of the Project Management workflow is to ensure the successful management of your project team's efforts. This includes both the technical aspects of project management — planning, scheduling, estimating, project viability assessment, metrics management, subcontractor management, and risk management — as well as the softer aspects such as people management and navigating your organization's political waters.

There is far more to project management than planning, estimating, and scheduling.

What are some of the best practices that your team can adopt with respect to project management? First, you may find that you need to break everyone out of the serial mindset, including yourself, and put yourself into the iterative and incremental mindset of the Unified Process. This is a fundamental principle of the Unified Process and of modern software development in general — many managers, particularly oder ones, are still in a serial mindset. Second, focus your team on building software, not on creating documents. If an artifact doesn't directly contribute to the development or validation of software, then you must seriously question why it is being developed. A UML class model that your programmers can use as the basis to write code is of great value; a logical data model that you create simply so it can be checked off of some senior manager's task list isn't. Third, build your team and foster a team culture that promotes teamwork and the development of software that reflects your project goals. As you bring people onto your team, you should assess their abilities and career goals, map them to the needs of your project, and where needed, obtain training and education to fill any gaps. This training and education effort may include formal courses in specific technologies or techniques, computer-based training (CBT), mentoring, or even bag-lunch sessions. Fourth, as mentioned previously in Chapter 3, you must accept that change happens. The reality of software development is that it is an uncertain endeavor — requirements and your project vision may evolve over time, even who and how your system will be used can change during its development. Your project management processes must reflect this reality.

Do not try to manage through precise prediction and rigid control structures. — James A. Highsmith III

Finally, be realistic. Virtually every project manager is pressured to develop schedules and estimates, that are, well... incredibly optimistic. The reality is that software development is often chaotic, affected by factors outside of your control, such as changing management strategies and technologies. A project plan that is based on everything going your way, and being incredibly lucky on top of that, is the mark of a doomed effort. To understand some of the fundamental mistakes that project managers can make, we highly suggest the book *AntiPatterns in Project Management* (Brown, McCormick, & Thomas, 2000). Uunderstanding what can go wrong is the first step in avoiding problems. Along a similar line of thought, we start this chapter with Scott Ambler's "Debunking Object-Oriented Myths" (February 1999, Section 5.5.1), a discussion of the morass of misunderstandings that exists about object technology and techniques within the software development industry. These myths include:

- Object-oriented development is (purely) iterative.
- Object-oriented development requires less testing.
- Object-oriented development has a significantly different software process.
- You shouldn't use relational databases to store objects.
- The Unified Modeling Language (UML) is complete.
- You don't have to worry about portability with objects.
- You have to use objects to do component development.
- Component-based development supersedes object-oriented development.

5.1 Starting Out Right

Everyone has to start somewhere, even software development managers. In "A Project Management Primer" (June 1998, Section 5.5.2), Karl Wiegers sets a high moral tone, spelling out the basic virtues that effective software managers must master for long-term success and exhorting them to put first things first. Although addressed to the new manager, even those who have been through several projects will find this article to be a useful review of critical basics that can help you gauge progress and fill in the gaps within your project management repertoires. If nothing else, Wiegers describes best practices to establish yourself as an effective team leader by setting the right priorities from the start.

For everything, there is a first time. — Spock

5.2 Technical Project Management Activities

Although there is far more to project management than its technical aspects, the fact of the matter is that these aspects are still crucial. In this section we cover topics such as:

- Justifying your project
- Planning your project
- Managing project risk
- Managing web-based projects in web time
- Outsourcing and subcontractor management
- Managing your measurements efforts

5.2.1 Justifying Your Project

A *business case* (Kruchten, 2000) is an artifact that describes the economic justification of your project — important information that can be used to obtain funding for your project. Your business case will also summarize the business context, success criteria, financial forecast, initial risk assessment, and plan for your project. A fundamental part of a business case is the justification of your project. Figure 5.1 depicts the solution to the Justify process pattern (Ambler, 1998b), indicating that there are three aspects to justifying a project: determining its economic, technical, and operational feasibility. While this is happening, potential project risks are being identified. The main deliverables of this effort are a feasibility study and an updated risk assessment for the project — information that is used as part of your project's business case.

Figure 5.1   The solution to the Justify process pattern.

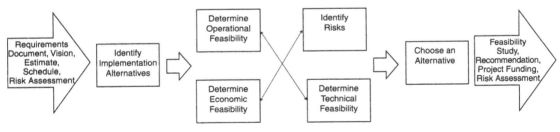

To justify a project, you need to perform a feasibility study which compares the various implementation alternatives based on their economic, technical, and operational feasibility. Based on the results of the study, you make a recommendation to accept one of the alternatives. The steps of creating a feasibility study are:

1. Determine implementation alternatives.
2. Assess the economic feasibility for each alternative.

3. Assess the technical feasibility for each alternative.
4. Assess the operational feasibility for each alternative.
5. Choose an alternative.

Table 5.1 lists some of the potential costs and benefits that may be accrued by a software project. Although the list is not comprehensive, it does provide an indication of the range of factors that you should take into consideration when assessing the economic feasibility of an application. The table includes both qualitative factors — costs or benefits that are subjective in nature, and quantitative factors — costs or benefits for which monetary values can easily be identified.

Table 5.1   Potential costs and benefits of a software project.

Type	Potential Costs	Potential Benefits
Quantitative	Hardware/software upgrades Fully-burdened cost of labor (salary + benefits) Support costs for the application Expected operational costs Training costs for users to learn the application Training costs to train developers in new/updated technologies	Reduced operating costs Reduced personnel costs from a reduction in staff Increased revenue from additional sales of your organizations products/services
Qualitative	Increased employee dissatisfaction from fear of change Negative public perception from layoffs as the result of automation	Improved decisions as the result of access to accurate and timely information Raising of existing — or introduction of a new — barrier to entry within your industry to keep competition out of your market Positive public perception that your organization is an innovator

Table 5.2 describes two basic categories of issues that should be addressed as part of your technical assessment. The first category addresses hard-core technology issues such as the scalability of the technology, whereas the second category addresses market issues such as the viability of the vendor.

Not only must an application make economic and technical sense, it must also make operational sense. The basic question that you are trying to answer is: "is it possible to maintain and support this application once it is in production?" Building an application is decidedly different than operating it, therefore, you need to determine whether or not you can effectively operate and support it. Table 5.3 provides examples of potential operational and support issues that you should consider when determining the operational feasibility of a system.

Table 5.2   Issues that should be addressed when assessing a technology.

Technology Issues	Market Issues
Performance Ease of learning Ease of deployment Ease of support Operational characteristics (i.e., can it run 24 hours a day, 7 days a week?) Inter-operability with your other key technologies Scaleability	Vendor viability (i.e., is it likely that they will be in business in two years? In five?) Alternate sources for the technology, if any Third-party support for related products and services Level of support provided by the vendor Industry mindshare of the product (i.e. is the market gravitating toward or away from this technology?)

Table 5.3   Issues to consider when determining the operational feasibility of a project.

Operations Issues	Support Issues
What tools are needed to support operations? What skills will operators need to be trained in? What processes need to be created and/or updated? What documentation does operations need?	What documentation will users be given? What training will users be given? How will change requests be managed?

During your justification efforts, you will identify and define many potential risks. It is critical that they are added to your risk assessment document so that they may be dealt with appropriately during the project. Some of the potential risks that you may identify are:

- the use of unproven (new and/or beta) technology on the project,
- waiting for vaporware (software products that have been promised but not yet released),
- inexperience of your organization with a given technology or process,
- nobody else has attempted a project of this size with the given technology,
- the use of several unfamiliar technologies or processes, and/or
- potential inability to support and operate the application once in production.

In "Mission Possible" (July 1996, Section 5.5.3), Jim Highsmith and Lynn Nix describe how feasibility studies can be applied to help you improve your chance of success. Feasibility studies answer three questions: (1) what is this project all about?, (2) should we do this project?, and (3) how should we go about this project?. Highsmith and Nix believe that a feasibility study is the place to nurture, examine, and question; don't have to take a long time; and need to be done the right way with the right attitude. You should identify and analyze alternatives. "To be, or not to be?" is the question a feasibility study should answer. The reality is that a staggering 31.1% of projects will be canceled before they are completed, and

52.7% will overrun their initial cost estimates by 189%. Had an attempt to justify the initial approaches to these projects happened, we would hope that some of these disasters could have been averted through the adoption of a better strategy.

To be or not to be, that is the question. — *William Shakespeare*

5.2.2 Planning Your Project

During the Inception phase, you need to develop what is called a coarse-grained project plan — effectively a high-level plan that identifies the major milestones and key deliverables of your team. This plan should also identify the initial iterations that you expect in each phase — iterations identified based on your project's requirements (the topic of Chapter 3). A common mistake that project managers make is attempting to develop an intricately detailed plan for an entire project at the very beginning — an unrealistic goal at best, and a recipe for disaster at worst. Although long-term, detailed planning is a noble goal, it is unrealistic and arguably misguided. At the very beginning of a projec, it isn't defined well enough to determine size, the architecture isn't in place so you do not know the technical factors involved, and you haven't identified the people that will be working so you have no way of estimating how effective your team will be. Planning exact tasks that you will perform several months in advance is as realistic as planning when you can take a 15-minute catnap several weeks from now — your situation is likely to change between now and then and force you to rework your plan.

A detailed, long-term project plan is a recipe for disaster.

Although it is very difficult to develop a detailed plan for the entire project, it is possible to develop one for the immediate time frame — namely, the Inception phase. An important part of project management during any project phase is the definition and continual update of a detailed project plan. Your planning efforts will include the definition of an estimate and a schedule — artifacts that you will update periodically throughout your project.

We have met the enemy and they is us. — *Pogo*

Joseph D. Launi, in "Creating a Project Plan" (May 1999, Section 5.5.4), provides descriptions of best practices to keep your project under control by balancing its scope, cost, time, and quality with stakeholder expectations. His experience is that a solid project plan is a blueprint that charts the entire project's course. It minimizes the cost of rework by anticipating and addressing problems early in the project. Although Launi discusses the project "diamond" in his article, a project "hexagon," as indicated in Figure 5.2, is likely more accurate (Ambler, 1998b). Six fundamental factors affect the estimate of a project: scope, cost, people, tools/techniques, quality, and time. The *scope* refers to the functionality that will be delivered by the application and *cost* refers to the amount of money invested to deliver it. The *people* assigned to a project affect the estimate because different people have different skillsets and levels of productivity. Similarly, the *tools* and *techniques* employed by the project team will provide varying levels of productivity. The level of *quality* that your organization is willing to accept also affects your estimate because the greater the desired quality, the more work that needs to be performed to achieve it. Finally, the desired delivery date of timeframe of the project has a major impact on your estimate because it will drive your decisions as to who will work on the project, what will be accomplished, and the tools/techniques you will employ. In this article, Launi combines pre-project discovery methods with proven management science techniques, consisting of the following steps in an iterative manner:

- Performing factor analysis
- Developing project agreement among stakeholders
- Identifying change control procedures
- Developing a work breakdown structure
- Estimating tasks
- Creating a schedule
- Performing a risk assessment

Figure 5.2   The iron hexagon of project planning.

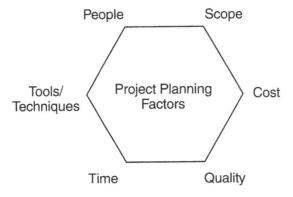

5.2.3 Managing Project Risk

A simple definition of a "risk" is a problem that hasn't happened yet, but could cause some loss or threaten the success of your project if it did. While you can never predict the future with certainty, you can apply risk management best practices to peek over the horizon at the traps that might be looming, and take actions to minimize the likelihood or impact of potential problems. Risk management means dealing with a concern before it becomes a crisis. It consists of several sub-activities including assessment, identification, analysis, prioritization, avoidance, control, management planning, resolution, and monitoring. In "Know Your Enemy: Software Risk Management" (December 1998, Section 5.5.5), Karl Wiegers, author of *Creating a Software Engineering Culture* (1996) and *Software Requirements* (1999), provides advice on how to improve the chance of successful project completion and reduces the consequences of risks that cannot be avoided. Wiegers, correctly so, believes that the list of bad things that can befall a software project is depressingly long — including external dependencies, requirements risks, management challenges, and lack of knowledge. An effective project manager should acquire extensive lists of these potential problems to help the team uncover as many risks as possible early in a software project.

The greatest risk we face in software development is that of overestimating our own knowledge. — James A. Highsmith III

5.2.4 Managing Web-Based Projects in Web-Time

Virtually every single organization in the world is either doing business on the Internet or at least is thinking about doing business on the Internet. The rate of change in the Internet and intranet development world is unprecedented, and as a result, methodologies are hard pressed to keep up. William Roetzheim, in the article "Estimating Internet Development" (August 2000, Section 5.5.6), describes how to accurately estimate Internet-based software. He points out that while traditional estimating has significant applicability to Internet projects there are some crucial differences. One issue is the fact that Internet software development projects now encompass activities that were previously done *prior* to the software project, particularly, market-strategy formulation and business-process analysis — activities that would be performed during the Inception phase. In this article, Roetzheim describes a process that yields:

- development and life-cycle costs,
- optimum staffing curves by labor category,
- costs broken down by phase and time,
- cost allocation to individual activities,
- a deliverable document list with descriptions and projected page counts, and
- a risk profile for the project.

If you think the release cycle has reached a frenetic and unsustainable pace in desktop software, wait until you hear what the World Wide Web has in store. How do you manage projects when the delivery date follows the initial budgeting by only a dozen weeks? How do you produce reliable, usable systems when there hardly seems time for coffee or a bathroom break? Dave Thomas, in "Web Time Software Development" (October 1998, Section 5.5.7), tells you how. One of the pioneers of object technology, Thomas and his handpicked crew have learned from a long string of successes how to manage development in "web time" and deliver good software on schedule. Not everyone will like what he has to say though. Thomas is one of those no-bull working managers who tells it like it is, without so much as a nod to trends, established opinion, or the answers people want to hear. And he speaks from experience. The reality is that to be competitive, companies must deliver products in web time (where a year is only three or four months long). Thomas describes best practices for developing "first-time" software; assessing technical risks; achieving the right level of quality; building just-in-time (JIT) software; effective use of guidelines, specifications, and reviews; component and assembly life cycles; managing by time boxes; for testing; release engineering; and managing by common peopleware principles.

You must deliver products in web time, not dinosaur time.

5.2.5 Outsourcing and Subcontractor Management

"Should I buy, build, or outsource?" is a question that should be asked by every single project manager during the Inception phase. You may decide to outsource all, or a portion, of your project efforts, and if so, what do you need to do to be successful? Steve McConnell answers this question in "Managing Outsourced Projects" (December 1997, Section 5.5.8). McConnell, the author of the Jolt-award winning *Rapid Development* (1996), believes outsourced projects force the buyer to practice good management. You need to manage an outsourced project at least as well as you manage your in-house development projects, but there are some added complications. To be successful, you must:

- specify requirements in detail,
- make an explicit decision to outsource,
- obtain resources,
- select a vendor to develop the system,
- write the contract, and
- implement in-house management and monitoring.

Neil Whitten provides complementary advice to McConnell's in *"Selecting the Best Vendor"* (November 1995, Section 5.5.9). He discusses how to develop an effective vendor contract, define the work that the vendor will perform, identify candidate vendors, and finally, manage the bidding process. He presents a collection of best practices, such as always seeking bids, ensuring that vendor candidates are qualified, preparing a complete and well-thought-out proposal, defining an objective bid proposal process, and not assuming anything when evaluating a proposal.

Select the best vendor that you can when outsourcing.

5.2.6 Managing Your Measurement Efforts

Metrics management is the process of collecting, summarizing, and acting on measurements (metrics) to potentially improve both the quality of the software that your organization develops and the actual software process. As the old saying goes, "you cannot improve it if you cannot measure it." Metrics management provides the information that is critical to understanding the quality of the work that your project team has performed and how effectively that work was performed. Metrics range from simple measurements of work effort, such as the number of hours invested to develop a model, to code quality measurements, such as the percentage of comments within your code.

If you can't measure it, you can't improve it.

Metrics management is a key part of software project management. In "A Software Metrics Primer" (Software Development, July 1999), Karl Wiegers, author of *Software Engineering Culture* (Wiegers, 1996) and *Software Requirements* (1999), points out that you can't track project status if you can't measure your overall effort to compare against your plan. You can't determine if your system is ready to deploy unless you know that your defect rate has been reduced to an acceptable level. You also can't determine how well your new development processes are working without metrics to act as a basis of comparison. The problem lies in which metrics should you collect. But luckily, Wiegers describes the Goal-Question-Metric (GQM) method to determine exactly that. He also points out that to be successful, you need to build a measurement culture with your teams and within your organization, and Wiegers provides advice for accomplishing that. Finally, he presents the following proven best practices for making metrics management a success within your organization:

- Start small.
- Explain why.
- Share the data.
- Define data items and procedures.
- Understand trends.
- Link overall organizational goals and process improvement efforts.

As Wiegers points out, metrics help you to manage and control your projects, and determine whether you are meeting your goals or missing them by a mile. Table 5.4 presents a wide range of potential metrics that you may decide to collect throughout your project. In fact, there are so many, that you need to narrow the list down to the handful that are important to you.

There are many metrics that you can collect; choose the ones that address issues that are important to you.

Table 5.4   Potential metrics to collect throughout the enhanced lifecycle for the Unified Process.

Workflow	Potential Metrics
Requirements	• Number of use cases • Function/feature points • Instability
Project Management	Level of risk Project size Cost/benefit breakpoint Number of lessons learned Percentage of staff members assessed/reviewed Calendar time expended Overtime Staff turnover Work effort expended
Infrastructure Management	Number of reused infrastructure artifacts Number of introduced infrastructure artifacts Number of candidate items for generalization Percentage of items generalized Effort required to generalize an item

Workflow	Potential Metrics
Analysis and Design	Method count of a class Number of instance attributes of a class Inheritance tree depth Number of children of a class Number of class attributes Number of ignored methods within a subclass
Implementation	Method size Method response Comments per method Percentage of commented methods Global usage Work effort to fix a defect
Test	Percentage of deliverables reviewed Time to fix defects Defect recurrence Defect type recurrence Defect severity count Defect source count Percentage of defects reworked
Deployment	Enhancements implemented per application release Problems closed per release Percentage of customers trained Average training time per person
Operations and Support	Average support request response time Average support request resolution time Support request volume Support request backlog Support request aging Support engineer efficiency Reopened support requests
Configuration and Change Management	Mean time between failures Software change requests (SCRs) opened and closed

Wiegers provides complementary advice in "Metrics: 10 Traps to Avoid" (October 1997, Section 5.5.11). By being aware of these common risks, you can chart a course toward successful measurement of your organization's software development activities. These risks, bordering on antipatterns, are:

- lack of management commitment,
- measuring too much, too soon,
- measuring too little, too late,
- measuring the wrong things,
- imprecise metrics definitions,
- using metrics data to evaluate individuals,

- using metrics to motivate rather than to understand,
- collecting unused data,
- lack of communication and training, and
- misinterpreting metrics data.

Warren Keuffel presents an interesting look at metrics applicable to object-oriented development in "Don't Fence Me In" (February 1995, Section 5.5.12). He argues that object orientation provides some interesting challenges for the individual attempting to apply engineering principles — principles that find their fundamental core in measurement. He presents an overview of many important object-oriented metrics, both from a theoretical and a practical perspective, and shares many lessons learned. Although this article is a few years old, it is still current — object-oriented metrics is a topic the software industry quickly came to a consensus about.

Often the biggest obstacle to metrics management is getting the developers — the ones in the trenches designing, coding, documenting and testing the software — to measure their work. In "Software Measurement: What's In It for Me?" (March 2000, Section 5.5.13) Arlene Minkiewicz presents a realistic look at how to be successful implementing a metrics program within your organization. She describes the steps that you need to take to ensure that the software designers, developers, testers, and others in your organization buy into your measurement program. From her experience, she believes that you should start by having clear goals for the measurement program, and then ensure that everyone on the team understands both what these goals are and what the organization expects to gain. You also need to follow up with periodic reviews of the measurement program to determine whether these goals are being met — recording the review results in concise and easy-to-understand terms. She also shares Karl Wieger's belief that you should never let your measurement program become a basis for individual performance evaluation — a mistake that is often the shortest path to metrics failure.

Developers must accept the need to collect metrics for your measurement program to succeed.

5.3 Soft Project Management Activities

A key activity of project management is to manage the people on the team — ensuring that they work together effectively and that the personal career goals of each team member are being met. People management is the act of organizing, monitoring, coaching, and motivating people in a manner that ensures they work together well and contribute to a project/organization positively. From our experience, one of the most important lessons that you can learn is that people aren't interchangeable parts. Every single person has a different set of skills, experiences, strengths, and weaknesses.

Good software comes from people. So does bad software.

When you stop to think about it, software developers are paid to solve problems, and in "Habits of Productive Problem Solvers" (August 1999, Section 5.5.14), Larry Constantine provides advice for doing so. An effective project manager understands that the best problem solvers learn the tacit logic of creative problem solving, whether they are conscious of the axioms and rules of this logic or not, and that they will often follow strategies that seem strange at first. Effective problem solvers avoid spinning their wheels by recognizing that they're doing so and then choosing to do something else — often starting with brainstorming or blankfilling. Some people, when presented with a tough problem, often choose to "sleep on it" in the hopes that they will awake the next morning with a new idea or fresh approach. The lesson to be gained from this article is that not only is there a wide range of techniques for solving problems, but that as a project manager, you must be prepared to manage people that will apply these techniques in their own unique manners.

Your management practices must reflect the fact that developers solve problems using a variety of techniques.

Managing developers can be hard enough when you are an experienced project manager, but what happens when you are new to the position? Andrew Downs presents advice for newly-minted technical leads in "From Engineer to Technical Lead" (April 2000, Section 5.5.15). In this new role, "soft" skills such as running meetings, communicating effectively, reviewing performance, and scheduling and reviewing code will be just as important as your technical background. Furthermore, if you leave the human issues unattended, they can lead to hard problems. To make matters worse, whether you have been promoted from within the organization or not, you'll face the challenge of earning your colleagues' respect. Downs presents best practices for starting off right, keeping everyone informed, to managing developers, growing with your team, and playing well with others. In short, this is an excellent article for any project manager to read, but particularly for anyone in a new leadership role.

During the Inception phase, the potential team members, including both system professionals and members of the user community, will be identified and recruited. You generally do not need to recruit everyone on the first day of your project — not that everyone is likely to be available anyway, but you must be prepared to meet the resource needs of your project throughout its lifecycle in a timely manner. You will find that you need to recruit within your organization for existing staff and perhaps externally as well for "new blood." Susan Glassett sets the stage in "Effective Resource Management" (September 1999, Section 5.5.16). She points out that people are a significant factor in the success of your project, and that by missing even one key person on a team, your project can easily run aground. It is a challenge to recruit and then retain the right people for your team — the better the match of a person to their position, the more productive they generally are. She correctly points out that people management goes beyond the scope of a single project, providing a collection of best practices for managing your staff effectively. Some of her suggestions include:

- Sett up a resource inventory.
- Record your staff's proficiency.
- Build project plans.
- Make assignments.
- Review your project teams and individual assignments.
- Perform capacity and contingency planning.

> **The market for systems professionals is global; pay your people what they are worth.**

A key responsibility of a project manager is to ensure that team members do not become innocent victims in the political battles and power struggles between senior managers and groups within your organization. Although we would like to advise you to isolate your staff from political issues, the fact remains that this goal is not possible and it is not even a very good idea. Savvy politicians will often pounce on your unsuspecting team members, asking them for information that they will later use against you. The best strategy is to make your staff aware of the political realities of your project — to identify who your friends and foes are, and then try to protect them as best you can from the politics so that they can concentrate on getting their jobs done.

> **Politics are a nasty reality of any software project. Ignore them at your peril.**

5.4 A Few More Thoughts...

Christine Comaford, in "What's Wrong with Software Development" (November 1995, Section 5.5.17), looks at common problems experienced during software development and presents a collection of best practices for solving them. She believes that developers need to:

- Better understand how to develop software.
- Better understand our users.
- Better manage our development staff.
- Create and maintain an application development infrastructure.
- Understand what rapid application development (RAD) really means.
- Produce reliable estimates.
- Control prototyping efforts better.
- Build better architectures.
- Test effectively.

To complement Comaford's insights, we end this chapter with Larry Constantine's "Scaling Up Management" (November 1998, Section 5.5.18), a look at the valuable lessons learned on large-scale development projects. Constantine argues that a "large" project is any project that is sufficiently bigger than anything you have done before, so that you have no idea how you are going to manage it. The question of scale is partly a matter of numbers, such as the number of people participating in a project, but a large part of the question is really about complexity. Why are large projects important? If you look at the big picture, it can help you manage a software project of almost any size. This article reviews the lessons learned during the making of Boeing's 777, created by a team of 10,000 people scattered around the globe. Insiders describe the 777 as two-and-a-half million lines of Ada code flying in close formation — not so much an airplane as a complex software system packaged and shipped in an airframe. Their approach to building the 777 included many interesting aspects, but the pieces that seem particularly relevant to complex software projects were the need for face-to-face team building, consensus-based collaboration, managing interfaces between teams through models, and reusing existing designs. Best practices that projects of any size can benefit from.

A large project is anything significantly bigger than you have done before.

5.5 The Articles

5.5.1 "Debunking Object-Oriented Myths"

by Scott Ambler

A morass of object technology myths exist within the software development industry. Learning the truth can save you from blindly falling into serious trouble.

There are many persistent object-oriented development myths, some disproved years ago, in the computer industry. Further, with the advent of new technologies and their marketing campaigns, new myths appear regularly. This month, I'll explore eight common myths, discuss their sources, and then address what I believe to be the truth. This column should help you make better technical decisions about object development.

Myth #1: Object-oriented development is iterative.

This is my favorite myth, one that I believe will never leave our industry. This myth was popularized in the late 1980s when everyone was excited about iterative techniques such as rapid application development (RAD) and prototyping. Most projects using object technology were small, proof-of-concept efforts that lasted a few months at best and focused mostly on construction efforts. Because these projects all but ignored the project initiation and deployment efforts of larger projects, many people thought object development must be iterative. The plethora of articles and books published in the early 1990s that focused solely on modeling and programming rather than deployment and initiation didn't help the issue either.

I think mission-critical projects using object technology are serial in the large scale, iterative in the small scale, and deliver incremental releases over time. This is because you will spend significant time at the beginning of a project getting organized, mostly planning and obtaining resources. You will then spend the bulk of your efforts performing construction-related activities such as modeling, programming, and testing in the small scale. Eventually you baseline your work, test it in the large scale, and release it to your user community. Once this is complete, you support your software and enhance it over time. It is clear from this that object development is serial in the large scale. However, object development is iterative in the small scale because on any given day you could be modeling, programming, testing, and planning.

At first it's difficult to grasp that object development isn't purely iterative, so I invite you to look at the three leading object-oriented software processes: the Unified Process, the OPEN Process, and the process patterns of the object-oriented software process (OOSP). You'll see that the four serial phases of the Unified Process (Inception, Elaboration, Construction, and Transition) are similar to the first three phase process patterns of the OOSP shown in Figure 5.3, although the Unified process doesn't include the OOSP's Maintain and Support phase. The contract-driven life cycle of the OPEN Process is depicted more iteratively than the other two, but it also takes a serial approach in large-scale projects.

Figure 5.3   The Object-Oriented Software Process (OOSP).

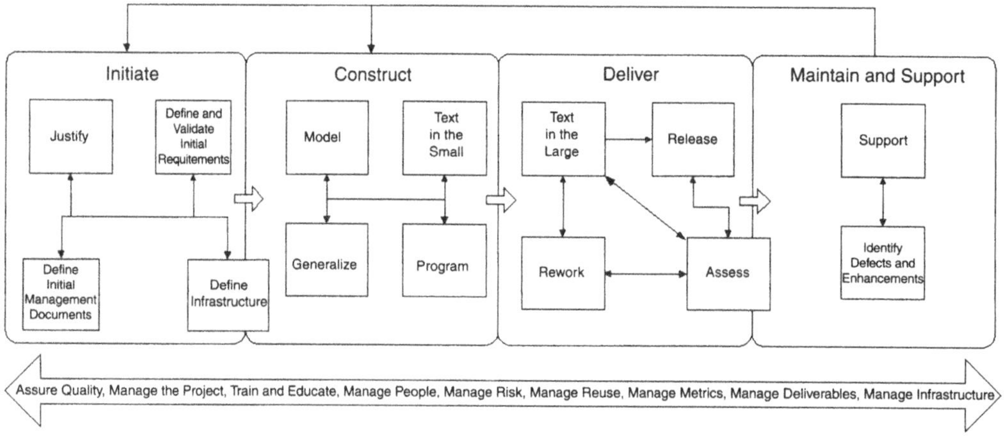

"Serial in the large, iterative in the small, delivering incremental releases over time."

Myth #2: Object-oriented development requires less testing.

Another prevalent myth in our industry is that you do less testing with objects. Nothing could be further from the truth. The reason why you use object-oriented techniques is because you want to develop software that is more complex than what you would develop with other approaches. More complex software that requires less testing? Stop and think about that for a minute.

This myth emerged from an assumption about inheritance: because your subclasses inherit the behavior of your superclasses, you don't need to rerun the tests of the superclasses when you test them. Yikes. Practice shows that you do need to retest, a technique called inheritance regression testing. If the tests run perfectly then that's great, but if they don't, then it was a good idea that you tested.

The reality is that good design and encapsulation will potentially lead to less testing of your software, and you can do both of these things regardless of the development paradigm you follow. You should be doing good design and encapsulating behavior with objects, but it's not a given.

Myth #3: Object-oriented development has a significantly different software process.

When object orientation emerged as a new paradigm, everyone assumed a new software process was needed. New techniques offer new ways to do things, but do we need a completely new software process? Seems unlikely to me. The reality is that, regardless of the development paradigm, you still need to do quality assurance, risk management, project management, requirements gathering, modeling, programming, testing, deployment, training, infrastructure management, maintenance, and support (to name a few). This was true with the procedural paradigm, it's true of the object paradigm, and it's true of the component paradigm. The general processes are the same, albeit some detailed techniques and procedures may be different. The bottom line is that the fundamentals of software development haven't changed.

Myth #4: You shouldn't use relational databases to store objects.

As many readers know, this is another one of my favorite myths. In the early 1990s, many object database vendors started up. A common theme of their marketing messages was that you couldn't use relational databases to store objects. Yes, at the time it wasn't completely clear how to map objects to relational databases, but that didn't mean it couldn't be done. Over time, we discovered that although relational databases aren't ideal for storing objects, they still work well enough. Today, object orientation is the dominant development paradigm, yet the relational database market is roughly $7 billion and the object database market is still only several hundred million dollars. The figures don't lie — storing objects in relational databases is the norm, not the exception.

Myth #5: The Unified Modeling Language (UML) is complete.

The UML has become the standard notation for object-oriented modeling. Because it is the standard, many people believe it's complete — which is surprising because we all know that standards evolve over time. In practice, developers quickly find that there are several models missing from the UML. For example, in The Unified Process Construction Phase (Ambler and

Constantine, 2000b) I argue that the UML was missing a persistence (data) model, an important model considering most people use relational databases to store their objects. The reality is that the UML is a good start, but it's still evolving.

Myth #6: You don't have to worry about portability with objects.

This myth is crazy, but unfortunately prevalent within our industry. It originates from Microsoft's marketing message that performance is more important than portability, so you need to use its development tools that provide good run-time performance — often at the expense of portability and maintainability (also locking you into its environment at the same time). Microsoft isn't the only culprit here. Relational database vendors do the same thing with their proprietary flavors of structured query language (SQL) and stored procedures (although kudos to Microsoft for introducing technology such as ODBC and OLE-DB to combat some of this).

Performance is important, but you will always have to worry about portability. Operating systems and databases change, even if they're only upgraded. Remember how much fun it was to port from Win16 to Win32? Have you ever had to port between databases? The need for portability is a reality in our industry, even for the minority of organizations that have a homogenous technology base. Object technology lets you write software that is easy to port — but only if you choose to do so.

Myth #7: You have to use objects to do component development.

Although this is a myth I can live with, the reality is that components and objects are orthogonal. This myth came about because objects and components work well together, and both are based on good engineering principles such as encapsulation, coupling, and cohesion. I've worked on projects that used procedural technology, including the use of stored procedures, to effectively implement large-scale domain components. I've also used Common Object Request Broker Architecture (CORBA)-based technologies to do the same. The reality is that component development can be performed using a wide range of tools and techniques, not just objects.

Myth #8: Component-based development supersedes object-oriented development.

This myth is partially true. Component-based approaches did appear after object-oriented approaches. Regardless of the numerous articles claiming that objects are dead and components are now number one, the reality is that object-orientation is the dominant paradigm and objects are being deployed in both component and non-component software. The main difference between components and objects is that components are typically implemented on a larger scale than objects. For example, in a bank, you may see a customer service component that handles all types of customer requests as well as a bank account object that handles transactions posted against one account. In this case, the customer service component would be built from many interacting objects. Just like the procedural paradigm didn't go away when the object paradigm came along, the component paradigm won't replace the object paradigm.

In this column, you saw that there is a morass of object technology myths within our industry. Developers who blindly believe them are destined to get into serious trouble. On the other hand, you've been lucky enough to have these myths exposed to you. Now comes the hard part — do you share this column with your colleagues or do you keep it to yourself? Some developers might call teamwork another myth within our industry.

5.5.2 "A Project Management Primer"

by Karl Wiegers, edited by Larry Constantine

New to project management? Establish yourself as an effective team leader by setting the right priorties from the start.

Everyone has to start somewhere, even software development managers. You may have approached the prospect with anxiety or unalloyed anticipation, but one day you found yourself "promoted" from the engineering or programming staff to a project lead or team lead position. Whether management is your chosen career path or you have only reluctantly agreed to try it, you've probably received little education in the arts of project and people management. In those first anxious days, it would be handy to have a primer to help you put first things first and keep you focused on actions that will improve both your own effectiveness and that of your team.

Set Priorities

First, you need to set your priorities as a manager. While you may be tempted to remain heavily engaged in software development activities, which may still be in your job description, you now have a new set of responsibilities. Many new managers cannot resist the lure of staying technically active, which can lead them to neglect the needs of others on the project team who look to them for help.

Effective leaders know their top priority is to provide services to their team. These services include coaching and mentoring, resolving problems and conflicts, providing resources, setting project goals and priorities, and providing technical guidance when appropriate. Team members need to know you are always available to help them. I find it valuable to think of myself as working for the people I supervise, not the reverse. Team members who need something from you are, in programming terms, a non-maskable interrupt over most anything else you might be doing.

Your second priority is to satisfy your organization's customers. As a manager, you no longer personally provide the products and services that satisfy customers. Instead, you must create an environment that enables your team to meet the needs of customers effectively.

Your third priority should be to work on your own projects. These may be technical projects or activities requested by your managers. Be prepared to postpone these activities when they conflict with the two higher priorities.

Explicitly taking actions to please your own managers should be your lowest priority. Unless you work in a Dilbert-style organization, your managers will be thrilled if you are successful at the three more important priorities. Even if you are not this fortunate, at least strive to keep the most important responsibilities of your new job at the top of your list. Instead of going out of your way to satisfy those above you, focus on helping your team members be as effective — and as happy — as possible.

Into the Gap

Regardless of your preparation, you may perceive some gaps in your current leadership and management skills. Your strong technical background was probably a factor in your being selected for the leadership role, but you'll need some additional skills to be fully effective. Take an honest inventory of your strengths and shortcomings in critical people and project skills, and begin closing any gaps.

Software developers aren't usually noted for their superlative people skills. You may need to become more adroit at interpersonal relationships, conflict resolution, persuasion, and selling ideas. You'll have to be able to deal with varied situations ranging from hiring and firing staff, to negotiating schedules, to comforting someone crying in your office during a performance review.

As technical team members, we may have enjoyed the luxury of energetically pushing our own agendas. However, effective management often requires a more collaborative and receptive interpersonal style. It took me a while to learn how and when to channel my natural assertiveness. I found it valuable to start my management career with a listening skills class. Effective listening includes being attentive, asking questions to build understanding, and responding supportively. What I learned has been useful in many situations.

Next, you may need to step to the other side of the podium and improve your presentation skills. If you are really uncomfortable with public speaking, a Dale Carnegie course might be helpful. Practice what you learn through such training, and you will find that your enhanced communication ability will serve you well in any job you hold in the future. I usually remind myself that I probably know more about the topic than the audience does, and that they're probably not hostile. Prepare your message, but don't recite it from memory. The audience doesn't know what you're going to say next, so don't sweat it if it doesn't come out exactly as you planned.

As a project leader, you will be responsible for coordinating the work of others, planning and tracking projects, and taking corrective actions when necessary to get a project back on track. Take a training course in project management and begin reading books and articles on project and risk management. Join the Project Management Institute (www.pmi.org) and read its monthly magazine, *PM Network*. The Software Engineering Institute's Software Capability Maturity Model (CMM), titled *The Capability Maturity Model: Guidelines for Improving the Software Process* (Addison-Wesley, 1995), contains useful advice on software project planning and project tracking. Your ability to set priorities, conduct effective meetings, and communicate clearly will have a substantial impact on your effectiveness as a manager.

Quality by Any Name

Almost everyone takes quality seriously and wants to produce high-quality products. However, there is no universal definition of what quality means in software. Debates rage about "good enough" software vs. more orthodox views of software quality. To help steer your group toward success, spend some time working with your team members and your customers to understand what quality means to them.

These two communities often do not have the same definition in mind, so it's easy to end up working at cross-purposes. A manager focused on the delivery schedule may be impatient with an engineer who wants to formally inspect every line of code. A customer to whom reliability is paramount won't be happy receiving a product that is piled with seldom-used features and riddled with bugs. A spiffy new GUI might turn off a user whose fingers have memorized how to use the previous version of the product most efficiently.

To better understand their views of software quality, my group at Kodak invited our customers (who were fellow employees) to an open forum on quality. This forum showed where our group's ideas of quality did not match the perceptions of those who used our products. Understanding such differences can help you focus energy where it will yield the greatest customer benefit, not just where it will provide the greatest developer satisfaction.

Traditional interpretations of software quality include conformance to specifications, satisfaction of customer needs, and the absence of defects in code and documentation. The buzzword of "six-sigma quality" may set a very high bar for low defect density or frequency of failure, but it doesn't address other dimensions of quality such as timeliness, usability, rich feature sets, and delivered value for the price. We might hope to maximize all of these characteristics in the products we produce and purchase, but trade-offs are always necessary.

During the requirements phase on one project, we listed 10 quality attributes we thought would be important to users. These included efficiency, interoperability, correctness, and ease of learning. We asked a group of key customer representatives to rate the desirability of each attribute. Once we understood which ones were most significant, we could design the application to achieve those objectives. If you do not learn what quality means to your customers and then design to deliver that level of quality, you're simply trusting your luck.

One telling indicator of high quality is that the customer comes back and the product does not. Work with your customers and developers to define appropriate quality goals for each product. Once determined, make these quality objectives unambiguous priorities. Lead by example and set high personal standards for the quality of your own work. You might adopt the motto: "Strive for perfection, but settle for excellence."

Recognize Progress

Recognizing and rewarding the achievements of your team members is an important way to keep them motivated. Unless your group already has a recognition program in place, this should be one of your top priorities. Recognition can range from the symbolic (certificates, traveling trophies) to the tangible (movie coupons, restaurant gift certificates, cash bonuses). Presenting recognition of some kind says, "Thanks for what you did to help," or "Congratulations on reaching that milestone." By investing a small amount of thought and money in a recognition and reward program, you can buy a lot of goodwill and future cooperation. Remember to recognize people outside the development group too, including customer representatives and support people who contribute in special ways to the project's success.

Talk to your team members to understand the sorts of recognition and rewards they find meaningful. Make recognition — for accomplishments large and small — a cornerstone of your team culture. Equally important is the implicit recognition of showing sincere interest in the work being done by each team member and doing all you can to remove obstacles to their effectiveness. Recognition is one way to demonstrate to your team members that you are aware of and appreciate the contributions they make.

Learn From the Past

It's possible that some of your group's past projects were not completely successful. Even on successful projects, we can often identify things we would do differently the next time. As you embark on your new leadership role, take some time to understand why earlier projects have struggled, and plan to avoid repeating the same mistakes. Software development is too hard for each manager to take the time to make every possible mistake on his or her own. Jump-start your own success by learning from what has worked (or not) before.

Begin with a nonjudgmental assessment, something that Norm Kerth calls a "retrospective", of the last few projects undertaken by your group, successful or not. Your goal is not to determine blame, but to do a better job on future projects. Conduct a post-project review to learn what went well and what could have been done better. Lead the team in brainstorming sessions or use an impartial facilitator to analyze each current project in the same way at major milestones.

In addition, become well-acquainted with established software industry best practices. A good place to start is with Part III of Steve McConnell's Jolt Award-winning *Rapid Development* (Microsoft Press, 1996), which describes 27 best practices. Beware of repeating the 36 classic software development mistakes he describes. Your team members may resist new ways of working, but your role is to ensure they consistently apply the best available methods, processes, and tools. Actively facilitate the sharing of information among team members so local best practices can become a part of every developer's toolkit.

Improvement Goals

Once you've conducted a retrospective into previous projects and determined what "quality" means to your group, set some goals for both short- and long-term improvements. Goals should be quantified whenever possible, so you can select a few simple metrics that will indicate whether you are making adequate progress.

For example, if you've determined that projects are often late because of volatile requirements, you might set a goal to improve the requirements stability by 50% within six months. Such a goal requires you to count the number of requirements changes per week or month, understand their origins, and take actions to control those changes. This will likely require alterations in the way you interact with those who supply the requirements changes.

Your goals and metrics make up part of the software process improvement program you should put into place. It's fashionable these days to disdain "process" as the last refuge of uncreative bureaucrats. The reality, though, is that every group can find ways to improve its work. Indeed, if you continue to work the way you always have, you shouldn't expect to achieve improved results.

There are two compelling reasons to improve your processes: correcting problems and preventing problems. Make sure your improvement efforts align with known or anticipated threats to project success. Lead your team in an exploration of the strengths and shortcomings of the practices currently being used, and of the risks facing your projects.

My group held a two-session brainstorming exercise to identify barriers to improving our software productivity and quality. In session one, the participants wrote their thoughts on sticky notes, one idea per note. A facilitator collected and grouped the ideas as they were generated. At the end, we had a dozen major categories, which we then recorded on large flip-chart sheets.

In the second session, the same participants wrote ideas for overcoming these barriers on sticky notes and attached them to the appropriate flipcharts. Further refinement led to a handful of specific action items that could be addressed in our effort to break down barriers and help team members achieve their software quality and productivity objectives.

Setting measurable and attainable goals brings focus to your improvement efforts. Keep the goals a visible priority, and monitor progress with the group. Remember that your objective is to improve the technical and business success achieved by your projects and company, not to satisfy the detailed expectations found in some process improvement book. Treat improvement efforts as mini-projects, with deliverables, resources, schedules, and accountability. Otherwise, process improvement activities will always get a lower priority than technical work, which you may find more enticing.

Start Slowly

Buffeted by the day-to-day pressures of managing, it can be a struggle just to keep your head above water. However, during this window of opportunity, the new manager has a critical role to play. You can't follow all the suggestions in this primer at once, but starting with what is most appropriate for your situation can help shape the culture and practices of your software development group for the long term.

Of course, completing the next project on time and within budget is vital, but as a software manager, you are responsible for doing more. You need to foster an environment of collaborative teamwork, lead the technical staff into a cohesive team that shares a commitment to quality, and promote and reward the application of superior software engineering practices — all while balancing the needs of your customers, company, team members, and yourself.

Good luck with the new job!

5.5.3 "Mission Possible"

by Jim Highsmith and Lynn Nix

**Your mission, if you wish to accept it:
Do a feasibility study before your project self-destructs.**

Software project failure is endemic in our industry. According to an article written by Jim Johnson, published in *American Programmer* ("Creating Chaos," July 1995), a staggering 31.1% of projects will get canceled before they are ever completed, and 52.7% will overrun their initial cost estimates by 189%.

Everyone is in a hurry to start, either because management has already delayed six months in initiating the project and needs to make up for lost time, or some competitor has just struck, or...well, you can fill in the blank. Many projects are injected with a tremendous oversupply of testosterone at their start. "Full speed ahead! Damn the torpedoes! Don't be a wimp!" Three or four months into the project, however, someone finally realizes no one knows what it's all about, and reevaluation sets in.

More than one-third of challenged or impaired projects experienced time overruns of 100% to 200%, according to Johnson's figures. One of the major causes of both cost and time overruns, he states, is restarts. For every 100 projects that start, there are 94 restarts. Ninety four percent is an incredible figure! These numbers suggest that we lose $78 billion dollars per year on canceled information technology application development projects. One solution to this dilemma is the feasibility study.

Feasibility studies answer three questions: What is this project all about? Should we do this project? How should we go about this project? A concept is not a project. A concept is a beginning, a thing to be examined and nurtured into a full-blown project. A feasibility study is the place to nurture, examine, and question. Feasibility studies don't have to take a long time, they just need to be done the right way and with the right attitude.

What Is This Project All About?

To paraphrase Lewis Carrol, "If we don't know where we want to go, then any path will do." One primary reason for project restarts, or outright failure, is the lack of a project mission, which at this early point means a careful analysis of the problems or opportunities and their possible impact on the organization. Team members, customers, and other stakeholders need a good understanding of the project's fundamental components — goals, objectives, scope, problem statement, constraints, and vision.

A good test of whether or not a project is understood is to walk around and ask various participants what they think it's all about. The more complicated the answer, the more trouble the project is in. A crisp, business-oriented, nontechnical answer usually means the project's groundwork is well established. The answer could be what we refer to as a project objective statement: a short, concise, high-level summary of the project. For example, "To identify and deliver a production-ready, state-of-the-art loan servicing system to include online collections and accounting subsystems by March 31, 1997."

Should We Do This Project?

The second major question answered by a good feasibility study is whether or not the project should proceed. The very name "feasibility" indicates one possible outcome is not to proceed. A significant portion of the $78 billion loss on software projects comes from projects that should never have gotten past the feasibility stage, but got caught up in corporate egos and politics.

Once the problems and opportunities have been identified, the next task of the feasibility study is to define the criteria for an acceptable solution. Feasibility (acceptability) incorporates political, economic, technical, and organizational components. For example, if the senior vice president of manufacturing demands a particular project to be done, why spend weeks coming up with a detailed cost/benefit analysis? In this case, the "should" question is fairly easy to answer. It is more effective to spend the remaining time answering the other feasibility questions.

The second phase of answering the "should" question is to identify the alternatives and recommend one. The alternative of not continuing the project should always be thoroughly considered.

How Should We Go About This Project?

A good feasibility study says more than "do it." In addition to defining the project objectives and deciding whether or not to proceed, it provides a broad outline of how to proceed. This involves preparing an initial, high-level project plan that provides a gross project sizing, identifies major milestones, and estimates resource needs. A plan of action serves two purposes: it gives the follow-up team a direction, and it forces the feasibility study team into thinking about critical implementation issues.

Figure 5.4   Feasibility study process.

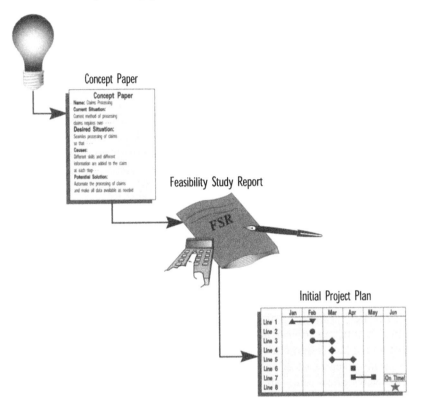

Concept Paper

Feasibility Study Report

Initial Project Plan

You can easily tailor a feasibility study to any organization, project, or development approach. The process is the same for all. The feasiblity study reveals what is required to build a solid business case, allowing management to make an informed decision about funding or canceling a project. We have oriented this article toward the feasibility of internal software projects. The process is similar for commercial software products, but the details would be tailored and more marketing-oriented.

Tips for a Sucessful Study

1. Understand the problem before jumping to a solution.
2. Always include key stakeholders in the feasibility process.
3. Carefully assess internal development capabilities.
4. Keep a balance between business and technical issues.
5. Define requirements clearly.
6. Distinguish the problem from the symptoms surrounding it.
7. Resolve political issues.

Organize the Team

Before you begin, organize your feasibility study team. How you structure this team will depend on your project and organizational structure. When choosing the team, ask yourself who the project stakeholders are and who represents each stakeholder.

It is crucial that you involve the stakeholders in the study, and it's even more important to correctly identify who these stakeholders are. You should have a sponsor involved who represents the client, directs the study, and approves the outcomes. A business analyst from the client organization and an information systems representative should also be part of this team. The business analyst and information systems representative will collaborate on all aspects of the study, with the information systems representative being the subject matter expert. All stakeholders should be identified, but identification alone isn't enough — they need to feel they are adequately represented during the study phase. This is done by actively soliciting their input and participation throughout.

Identify the Problem or Opportunity

A feasibility study is usually the response to some client-identified problem or opportunity. Clearly understanding the problem or opportunity provides the framework for all feasibility study activities. Go to the source — the individuals who identified the issue — to provide information regarding the origins and details of the problem or opportunity.

Avoid the tendency to jump to solutions without first carefully stating the problem. It takes discipline to keep asking, "Am I describing a solution instead of a problem? If so, what is the problem this solution solves? Is it really the root problem, or is it merely a symptom?" An example of a problem statement is: "Sales have increased to 4,000 orders per day on average with projected increases of 25% per year for the next five years. The current order processing system cycles allow only an average of 3,500 orders to be processed in a 24-hour period." Problems or opportunities should relate to increasing revenues, avoiding costs, improving customer satisfaction, complying with regulatory agencies, and improving the competitive situation. Any problem or opportunity not traceable to one of these benefits is suspect.

Interview all groups within the organization that will use the software to ensure the problem or opportunity has been defined from the global perspective. Joint application development sessions are helpful here. Problem analysis should include the following steps:

- Describe the current situation: What is happening now? What is the effect on the organization? What is the effect on your customers?

- Describe the desired situation: What should happen? How should employees be able to handle this situation? What would satisfy your customers? What should you do to be more competitive?

- Identify possible causes: What is forcing you to do business in the current manner? Why does the current system work the way it does?

- List potential solutions: What changes can you make to move toward the desired situation? Would the use of information technology help you to reach the desired situation? If so, describe the solution's general features.

Figure 5.5   Six simple steps for feasibility analysis.

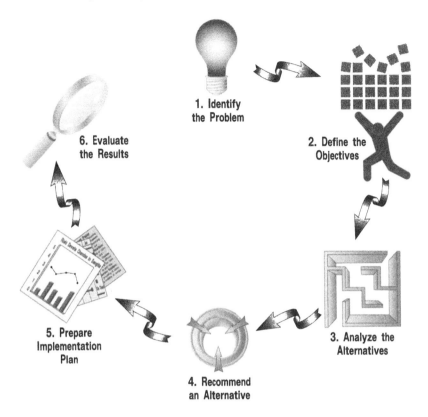

1. Identify
the Problem

6. Evaluate
the Results

2. Define the
Objectives

5. Prepare
Implementation
Plan

3. Analyze the
Alternatives

4. Recommend
an Alternative

Define the Project's Objectives and Criteria for Evaluating Solutions

Once the problems and opportunities are clear, you can write business and technical objectives. These will form the basis for evaluating alternative solutions. Separating business and technical objectives is important. Business objectives address one of the five areas (increasing revenue, and so forth) mentioned earlier. Technical objectives support business objectives. When identifying the business objectives, you must first ask which business results you are hoping to achieve.

For the objectives to be effective in evaluating alternatives and guiding the project, you must state them as clearly and measurably as possible. A clear objective contains the following components:

- Action: What will the solution do?
- Focus: In what areas will it do it?
- Quantity: By how much?
- Quality: With what level of achievement?
- Time: By when?

A business objective example is: "Increase competitiveness (target market share increase of 4%) and revenues (target increase of $35,000 per year), and improve customer satisfaction (reduce complaints by 30%) by implementing the new, two-day item delivery process by August. The process should be capable of handling up to 10,000 items per day, with an average of 7,000."

A technical objective supporting this business objective might be: "Implement a new, automated item processing application and associated client/server hardware and terminals to process up to 10,000 items per day with an average of 7,000 per day, and allow for a 10% annual growth rate."

The business organization is responsible for achieving the business objectives (revenue, and so forth) while the technical organization is only responsible for the technical objectives. Other evaluation criteria, such as risk, should also be added to the evaluation. Two risk areas often misjudged are complexity (business process, technical) and resources capability. While complexity is underestimated, organizations simultaneously overestimate their technical and project management capabilities — a deadly combination. One evaluation criteria might be: Can we implement this alternative with our current staff?

Identify and Analyze Alternatives

The highest-level alternatives are to do nothing, to maintain the current system, or to build a new system. Most often, however, analyzing the third option in more detail requires the most work. Will you develop the entire system? Will you use internal staff, outsource the project externally, or purchase the system components off the shelf? On which platforms (client/server, mainframe, the Internet) will the system run? As your team brainstorms possible solutions, the potential for real innovation and creativity arises.

Clearly describe each proposed solution and how well it meets the business and technical objectives. Define the scope of the solution, the high-level technical design approach, the time frame required to implement the solution, and any associated assumptions and constraints. To adequately accomplish this step, you will probably need to define preliminary requirements and high-level technical designs.

Ten Reasons "Not to be" (Signs of an Unfeasible Project)

1. Major political issues are unresolved by the feasibility study.
2. Key stakeholders won't participate in the feasibility study (and therefore probably the project).
3. Risks (probability of adverse consequences) are too high (technical, economic, organizational).
4. Cost and benefit ratio isn't favorable enough, especially when benefits are "soft."
5. Internal staff's experience and training is insufficient for the project.
6. Requirements are unclear, or keep changing radically during the feasibility study.
7. Risk and reward ratio is unfavorable. High risks usually need a high reward to be worthwhile.
8. Clients (in a multidisciplinary project) can't agree on exactly what the problems or objectives are.
9. No executive wants to be the project's sponsor.
10. Implementation planning seems superficial.

Analyze Each Alternative

Once you've identified all the possible ways of developing the solution, you must find the alternative that meets the largest number of business and technical objectives. A matrix can help. The matrix identifies the alternatives in rows along the left-hand side, the objectives to be met across the top, and appropriate rankings (money, weighted numbers) in the matrix cells. The rating should be money for quantifiable benefits and some weighted number for other factors or intangible benefits.

Next, do a cost and benefit analysis for each alternative. It is always easier to quantify costs than benefits. Even so, quantify benefits as precisely as possible (reducing invoice cycle time from 12 to 2 days, saving $35,000 per month in operating expenses, and so on). Well-thought-out benefits will help you sell the solution you propose. Intangible benefits (such as employee morale) should be quantified as well. Of course, you can't quantify an intangible, but you can use relative weighting factors or "soft" dollars. For example, you could rank alternatives on a scale of one to seven for how well they might improve customer satisfaction.

One component of project success, and an important evaluation criteria for alternative solutions, is risk. Analyze each alternative's functional, financial, organizational, and technical risk using a risk-evaluation checklist. Most risk analysis suffers from political timidity. People are reluctant to talk about risk because they might be interpreted as being pessimistic. Unfortunately, this reluctance dooms many projects to failure because the team will pick an overly risky alternative and then fail to actively manage those risks.

Recommend the Best Alternative

When doing a final analysis of your alternatives, you must evaluate each candidate for how well it fits within your organization's strategy, how it will affect your existing operations, what special equipment needs will be involved, and where funding for the solution will come from. Then you can forward your recommendation to the powers that be for approval and funding.

Give a frank assessment of the recommended alternative. Don't oversell. The sponsor won't trust the recommendation unless he or she can trust the analysis. Be rational and circumspect. Provide the sponsor with all rationale necessary for him or her to clearly understand why the one proposed alternative is the clear choice.

In the real world of problem solving there is often no clear winner. One solution may meet most of the objectives, but not all. Worse yet, there may be no alternative that adequately meets the objectives. Often, objectives will need to be revisited with the client, or additional information must be collected. Too often the technical staff becomes discouraged because the decision appears to be made on a political rather than analytical basis. The goal of the feasibility team should be to provide a comprehensive package of information about the project, not the decisions regarding the project itself. In fact, a decision to abandon the project should be viewed as a tremendous success — look how much money was saved!

Prepare an Implementation Plan

The last step of conducting a feasibility study is to develop a high-level implementation. As mentioned earlier, such a plan gives the follow-up team direction and forces the feasibility study team into thinking about critical implementation issues. The plan should identify implementation approaches such as: Will this be a more traditional or a rapid application development-style project? What equipment, facilities, and people will be required for implementation? Will the project use outside contract assistance? Actually, you will need to do some implementation planning to provide adequate cost information for alternative solutions.

Because software projects are learning journeys, we recommend using a phase-limited commitment approach where work estimates for the next milestone are developed in detail, while only general estimates are developed for subsequent milestones. As each milestone is reached, the remaining work is planned in more detail.

To Be, or Not To Be

"To be, or not to be?" that is the question a feasibility study answers. The success or failure of a project is often decided very early. To pull off an effective feasibility study, you must have the right attitude and the right approach. Having a good feasibility study process without the proper commitment from management and staff to listen to the answers doesn't work well — it results in substance without form. Having a commitment to listen, but without the substance of a reasonable feasibility study process isn't a lot better. Doing a feasibility study takes time up front, and will likely result in a later start date for a software project. The benefit you'll receive from starting slow, however, is a quality product finished on time and on budget.

5.5.4 "Creating a Project Plan"

by Joseph Launi

Keep your project under control by balancing its scope, cost, time, and quality with stakeholder expectations.

A solid project plan is a blueprint or a game plan that charts the entire project's course. It should minimize the cost of rework by anticipating and addressing problems early in the project. There is nothing more frustrating for a CIO than approving a project plan — only to find out months later that the recommended technology cannot be applied. To combat this, risk assessment is a mandatory step in effective project planning.

The fundamental premise of good project management states that the project manager's greatest challenge is effectively balancing the components of time, cost, scope, quality, and expectations. Figure 5.6 shows the project diamond, which signifies this balance.

The components of the project diamond have a symbiotic relationship. For example, when a user requests an additional report that wasn't agreed on in the requirement specifications, the project's scope and quality change. This will change the other project components as well. As a result, the diamond's shape will be skewed, graphically depicting a project out of control. The challenge is managing change while keeping the diamond's shape intact. Project planning defines the diamond, while effective change and expectation management lets you manage it throughout the project's life cycle.

Figure 5.6 The project diamond.

In this article, I'll define a methodical approach to assembling project plans. This approach will help your team manage client and upper management expectations by accurately and effectively defining any project's scope, cost, time, and quality dimensions. This process combines pre-project discovery methods with proven management science techniques, and consists of the following steps:

- Performing your factor analysis
- Developing your project agreement
- Creating your change control procedures
- Developing your work breakdown structure (WBS)
- Estimating the tasks
- Creating the schedule
- Performing risk assessment

Performing Your Factor Analysis

In the *Seven Habits of Highly Effective People* (Covey Leadership Center, 1998) author and management scientist Steve Covey writes, "To be understood, you must seek to understand." Often, from the moment a project is initiated, developers make assumptions about the effort ahead and how to attack it. Too often, they leap to conclusions about projects — to the point of preparing plans or estimates based largely on invalid or incomplete information — and the project fails to meet its budget, objective, and schedule expectations. Sooner or later (usually later) in the development process, the conditions you should have discovered in the planning process, like choosing the right mix of human resources, become apparent, and the project efforts — or deliverables — suffer. How many times have you completed a documentation task only to realize that a technical writer would have performed it better than the programmer?

Factor analysis is a disciplined technique for investigating, analyzing, and understanding a project — before you make commitments. Similar to requirements analysis in a standard development methodology, factor analysis questions should not only focus on system requirements but also the environmental issues surrounding the project. These issues include: What is the project's scope? Are all client personnel committed to the project? What is the acceptance criteria? How will I communicate project status? Often, a good factor analysis will lead to the inclusion of a technical requirements analysis phase in the system develop-ment effort.

This process consists of 10 key factors or elements of every project. Once you understand them, you and your team can explore each element's impact on the project and the appropri-ate project deliverables. Performing a factor analysis removes much of the mystery from plan-ning and estimating future events. It is a structured, critical first step in confirming what you know and exposing what you don't know about a project.

Following are the 10 factors of every project. As you gather information about the specific factors, it's important to document expectations, constraints, and associated issues to con-sider during subsequent planning activities.

Definition/Scope: The primary project's purpose, including major functions and deliver-ables, and the purpose of the project relative to the organizational whole

Resources: The financial, technical, material, and human resources you need to support or perform the project

Time: Elapsed time and actual work time required to complete the project

Procedures: The various organizational requirements: policies and procedures, methodolo-gies, quality program, work request initiation, financial review, and so on

Environment: The project's entire cultural, political, geographical, physical, and technical context

Change: New or different future conditions, requirements, events, or constraints discov-ered within or imposed on the project — and the associated procedures for managing project change

Communications: Meetings, status reports, presentations, and complexities that will affect communication

Commitment: The degree of sponsor, user, and other stakeholder support

Expectations: Business, productivity, operational, cultural, or any other performance changes you expect as a result of project implementation

Risk: The potential jeopardies to project success and, conversely, the jeopardies associated with not doing the project or some aspect of the project.

Develop Your Project Agreement

The project agreement is the contract or proposal that defines the work you need to do in terms of the project diamond. In addition, project agreements document the project schedule by depicting a work breakdown structure — the foundation for accurate estimates, task dependencies, and eventually a critical project path.

It is important to negotiate project agreements so that the client and the project manager agree on the work needed and the schedule to accomplish it. Stakeholders should understand their requirements, but they might not grasp the time frames and cost for development and implementation. The project manager must be comfortable debating all of the diamond's components. The ultimate goal should be a mutual understanding of the diamond's dimen-sions. This creates buy-in on both sides.

Creating Change Control Procedures

A project can change at any time. For example, new requirements can be introduced. Change management is simply the process of analyzing and evaluating potential changes in the project's scope, schedule, quality, or cost for orderly integration into the project plan. The project manager's role is not just to implement customer requirements but also to assist and facilitate in the impact assessment of all change items. The project manager must work with the client to make good business decisions. The change process, which you must document in the project agreement, is the vehicle to advance this.

A universal truth applies to any project: change will occur constantly, dynamically, and usually, without warning. "Freezing" requirements is a popular, though rarely feasible, approach to controlling the inevitable. A good project manager recognizes the inevitability of change and plans for effective change management.

Effective project management anticipates change and the calculated reaction to it. Ineffective change management will blur the boundaries of when and what is to be delivered. If you don't plan for change, the ground will be well-tilled for poisonous project weeds and skewed expectations.

Since my definition of quality is "conformance to customer requirements," the process of evaluating and communicating the effects of change with the client is fundamental to a strong project management process.

Developing Your Work Breakdown Structure

Creating a work breakdown structure is integral to planning, estimating, scheduling, and managing a project. It is the process of decomposing a project into its detailed work tasks, and logically arranging these tasks to accomplish the project's objectives. This includes identifying the phases, deliverables, and activities you need to accomplish project objectives.

The lowest level of a work breakdown structure provides the basis for estimating and cost-tracking. It's generally the most reliable level in which to estimate. More often than not, poor estimates reflect missing tasks, rather than poorly estimated, known tasks. Without a comprehensive work breakdown structure, you can't measure or predict project progress, costs, and accountability.

Moving down the hierarchy represents an increasing amount of work detail. The lowest detail level tasks are called work packages. Work packages are the units you schedule the project work by. You use the higher- or summary-level tasks for project sizing and negotiation, cost accounting, and management and customer status reporting. You also define project milestones in the work breakdown structure. Milestones are major events or points in time that you'll measure your progress against.

Figure 5.7 depicts a work breakdown, or decomposition, diagram. It shows the typical work breakdown structure numbering convention.

Figure 5.7   **Work breakdown structure.**

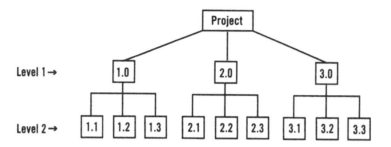

Estimating the Tasks

You must estimate the approximate time and costs you need to complete the project you described in the work breakdown structure, and in accordance with environmental constraints you identified during factor analysis. Preparing estimates of time and cost can be difficult for project managers. If you follow factor analysis and work breakdown structure principles properly, you'll develop estimates with greater confidence. Most estimate problems are caused by failing to identify all the necessary tasks, not from poor time estimation of the identified tasks. You cannot estimate time or costs for tasks that have not been identified.

You can develop estimates at various levels of detail. You can develop planning approximations for budgetary purposes with higher-level work breakdown structures than contract estimates for billing. In general, estimates prepared at the lowest level of the work breakdown structure are more reliable and accurate than estimates prepared at the higher levels.

Project managers often have problems because they don't compensate for a few basic estimating concepts. For example, it may take 40 hours to complete a programming task. This does not necessarily mean the task duration will last one week. If a programmer is in high demand and divided between two projects, this task could actually take two weeks. If a vacation week falls in those two weeks, it may take three weeks to complete this 40-hour task.

The time you spend estimating a project may be a small percentage of the total project time. It is still important to document the estimate, because it not only lets you improve the standard (for future estimates), it also lets you teach the estimating process to someone else.

Creating the Schedule

Once you identify all the project tasks and establish their duration, you can assign staff resources, derive task dependencies, and then develop the subsequent schedule. Scheduling is the process of varying scope, time, resource, and risk constraints to select and document the optimum project time and staff schedule. The project schedule will be your primary tool for monitoring the project life cycle's progress. You must revise it periodically to reflect the inevitable changes I discussed earlier.

Scheduling, like estimating, occurs throughout the project life cycle. First, create the initial schedules to negotiate the project agreement. Then, each time the project changes, you complete the work breakdown structure and associated estimates and rework the schedule. This includes analyzing the tasks and resource requirements to determine their impact on the remaining project.

Once you assemble the schedule, you should clearly understand the critical path. The project's critical path will determine the project completion date. The critical path has the least amount of float (or 0 float days) of all paths, and has the longest elapsed time of all paths.

In summary, any task on the critical path that slips its completion date will make the project's end date slip as well. Thus, tracking the critical path will help you determine whether or not your project will end on time. Equally important, tasks on the critical path may not be the most "project critical tasks." Critical path is a management science concept that only addresses the time dimension of the diamond. Contrary to popular belief, not all projects are time critical. The critical path may not be the place where you should spend the majority of your time. Critical path tasks, for example, might be documentation activities, while design and programming fall on paths with float. You may choose to spend the bulk of your time monitoring the design and programming activities while letting the documentation specialists inform you of any potential time delays.

Perform Risk Assessment

Risk is an inherent component of project management. Whenever you undertake any project, you expose yourself to risk. Project risk is anything that will adversely affect project objectives. Risk is the possibility of incurring some harm and therefore is related to the potential occurrence of some events. When assessing risk, consider these three factors:

- The risk event itself.
- The event's probability.
- The effect's potential severity.

Risk management is the process of identifying, analyzing, and responding to risk throughout the project's life cycle. This response should take the form of contingency plans — additional tasks built into the work breakdown structure to mitigate a risk event, deflecting the risk items to another entity, and so on. For example, a project team is developing a system that calls for the implementation of new technology. The team has the resources necessary to develop the system but fears it lacks the resources to implement the new technology. The risk scenario would look something like Table 5.5.

Table 5.5   Risk scenario examples.

Risk Event:	What if we don't have the resources to implement the new technology when it's needed late in the project?
Risk Probability:	75% chance the resources will not be available. 90% chance the resources, if available, will perform poorly.
Impact:	Project will not complete on time, quality of the final project will be compromised if the technology is not installed properly.
Response:	Find a qualified subcontractor to assist or perform the new technology implementation.

As you can see, the response in Table 5.5 will have a major impact on your project agreement. This adds greater complexity to the project and affects the cost by adding tasks to the work breakdown structure. In addition, it will affect the quality of the system you install. At a minimum, you should use this type of analysis for all risk events with a high probability of occurrence and a high degree of impact.

Manage the Diamond by Managing Expectation

Effective project planning is not conducted in a vacuum. It must be carried out in coordination and cooperation with all appropriate stakeholders. You must manage their expectations throughout the process. The project manager must constantly look for opportunities to create win-win relationships by negotiating the work that must be accomplished. A project manager who declares that "this can't be done in the time frame allotted" will meet with stiff resistance from client management. On the other hand, a project manager who can defend this statement with a solid understanding of the project's scope, backed by a logical work breakdown structure; thoughtful estimate; thoughtful estimate and project schedule; and concise risk analysis will be met with a response like, "Maybe you're right. Help me understand what you understand." This is effective expectation management and proper development of win-win relationships. Once your project plan is in place, it's easy to simply manage your project diamond.

5.5.5 "Know Your Enemy: Software Risk Management"
by Karl Wiegers

Structured risk management processes can help you discover potential problems with a project — and eliminate them before it's too late.

Software engineers are eternal optimists. When planning software projects, they often assume everything will go exactly as planned. Or, they take the other extreme position: the creative nature of software development means you can never predict what's going to happen, so what's the point of making detailed plans? Both of these perspectives can lead to software surprises when unexpected things happen that throw the project off track. In my experience, software surprises are never good news.

Risk management is becoming recognized as a best practice for reducing the surprise factor. While you can never predict the future with certainty, you can apply structured risk management practices to peek over the horizon at the traps that might be looming, and take actions to minimize the likelihood or impact of potential problems. Risk management means dealing with a concern before it becomes a crisis. This improves the chance of successful project completion and reduces the consequences of risks that cannot be avoided.

Why Manage Risks?

A simple definition of a "risk" is a problem that hasn't happened yet but could cause some loss or threaten the success of your project if it did. These potential problems might have an adverse impact on the cost, schedule, or technical success of the project; the quality of products; or team morale. Risk management is the process of identifying, addressing, and eliminating these potential problems before they can do damage.

A formal risk management process benefits both the project team and the organization as a whole. First, it gives you a structured way to evaluate threats to your project's success. By considering the potential impact of each risk item, you can focus on controlling the most severe risks first. You can combine risk assessment with project estimation to quantify possible schedule slippage if certain risks materialize into problems, thereby coming up with sensible contingency buffers. Sharing with your team what does and does not work to control risks across multiple projects helps avoid repeating past mistakes. Without a formal approach, you cannot ensure your risk management actions will be initiated in a timely fashion, completed as planned, and effective.

Controlling risks has a cost, which you must balance against the potential cost you could incur if the risk is not addressed and does indeed bite. For example, if you are concerned about a subcontractor's ability to deliver an essential component on time, you could engage multiple subcontractors to increase the chance that at least one will come through on schedule. That's an expensive remedy for a problem that may not even exist. Is it worth it? It depends on the down side you incur if the subcontractor slips. Only you can decide what's right for each situation.

Typical Software Risks

The list of evil things that can befall a software project is depressingly long. The enlightened project manager will acquire extensive lists of these risk categories to help the team uncover as many concerns as possible early in the planning process. Possible risks to consider can come from group brainstorming activities or from a risk factor chart accumulated from previous projects. One group I've worked with had individual team members come up with insightful descriptions of risk factors. I edited the factors together, and then we reviewed them as a team.

The Software Engineering Institute (SEI) has assembled a taxonomy of hierarchically organized risks in 13 major categories, with about 200 thought-provoking questions to help you spot the risks facing your project. These are listed in "Taxonomy-Based Risk Identification," by Marvin Carr, et al. (SEI Technical Report CMU/SEI-93-TR-006). Steve McConnell's Jolt Award-winning *Rapid Development* (Microsoft Press, 1996) is also an excellent resource on risk management.

Following are several typical risk categories and risk items that may threaten your project. Have any of these things happened to you? If so, add them to your master risk factor checklist, and ask future project managers if these things could happen to them, too. There are no magic solutions to any of these risk factors. You need to rely on your past experience and your knowledge of software engineering and management practices to control the risks that concern you most.

Dependencies

Many risks arise because of outside agencies or factors on which your project depends. You cannot usually control these external dependencies, so mitigation strategies may involve contingency plans to acquire a necessary component from a second source. Mitigation strategies could also include working with the source of the dependency to maintain good visibility into status and detect any looming problems. Here are some typical dependency-related risk factors:

- Customer-furnished items or information
- Internal and external subcontractor relationships

- Inter-component or inter-group dependencies
- Availability of trained, experienced people
- Reuse from one project to the next

Requirements

Many projects face uncertainty and turmoil around the product's requirements. While some of this uncertainty is tolerable in the early stages, it must be resolved as the project progresses. If you don't control requirements-related risk factors, you might either build the wrong product or build the right product badly. Either situation results in unpleasant surprises and unhappy customers. Watch out for these risk factors:

- Lack of a clear product vision
- Lack of agreement on product requirements
- Inadequate customer involvement in the requirements process
- Unprioritized requirements
- New market with uncertain needs
- Rapidly changing requirements
- Ineffective requirements change management process
- Inadequate impact analysis of requirements changes

Management

Don't be surprised if your risk management plan doesn't list many management shortcomings. After all, the project manager is usually the person who writes the risk management plan, and most people don't wish to air their own weaknesses (assuming they even recognize them) in public. Nonetheless, management risk factors inhibit the success of many projects. Defined project tracking processes and clear project roles and responsibilities can help address some common management risks:

- Inadequate planning and task identification
- Inadequate visibility into actual project status
- Unclear project ownership and decision making
- Unrealistic commitments made, sometimes for the wrong reasons
- Managers or customers with unrealistic expectations
- Staff personality conflicts

Lack of Knowledge

The rapid rate of change of software technologies, and the increasing shortage of talented staff, mean project teams may not have the skills they need to be successful. The key is to recognize the risk areas early enough so you can take appropriate preventive actions such as obtaining training, hiring consultants, and bringing the right people together on the project team. These factors might apply to your team:

- Lack of training
- Inadequate understanding of methods, tools, and techniques
- Insufficient application domain experience

- New technologies or development methods
- Ineffective, poorly documented, or ignored processes
- Technical approaches that may not work

Risk Management Approaches

Risk management consists of several sub-activities including assessment, identification, analysis, prioritization, avoidance, control, management planning, resolution, and monitoring.

Risk assessment is the action of examining a project and identifying areas of potential risk.

- *Risk identification* can be facilitated with the help of a checklist of common risk areas for software projects, or by examining the contents of an organizational database of previously identified risks and mitigation strategies (both successful and unsuccessful).
- *Risk analysis* involves examining how your project outcome might change with modification of risk input variables.
- *Risk prioritization* helps you focus the project on its most severe risks by assessing the risk exposure. Exposure is the product of two factors: the probability of incurring a loss due to the risk, and the potential magnitude of that loss. I usually estimate the probability from 0.1 (highly unlikely) to 1.0 (certain to happen), and the loss on a relative scale of 1 (no problemo) to 10 (deep tapioca). I multiply these factors to estimate the risk exposure due to each item, which can run from 0.1 (don't give it another thought) through 10 (stand back, here it comes!). It may be easier if you simply estimate both probability and impact as high, medium, or low. Those items having at least one dimension rated as high are the ones to worry about first.

Risk avoidance is one way you can deal with a risk: don't do the risky thing! You may avoid risks by not undertaking certain projects, or by relying on proven rather than cutting-edge technologies when possible.

Risk control is the process of managing risks to achieve desired outcomes.

- *Risk management* planning produces a plan for dealing with each significant risk, including mitigation approaches, owners, and timelines.
- *Risk resolution* is execution of the plans for dealing with each risk.
- *Risk monitoring* involves tracking your progress toward resolving each risk item.

Simply identifying the risks facing your project is not enough. You should also write them down in a way that lets you communicate the nature and status of risk factors throughout the affected stakeholder community during the project. Figure 5.8 shows a form I find convenient for documenting risks; you can also store this information in a spreadsheet. This risk list could be included as a section in your software project plan, or it could remain as a stand-alone document.

When documenting risk statements, use a condition-consequence format. That is, state the risk situation (condition) you're concerned about, followed by at least one potential adverse outcome (consequence) if that risk should turn into a problem. Often, people suggesting risks may state only the condition ("the customers don't agree on the product requirements") or the consequence ("we can only satisfy one of our major customers"). Pulling those together into the condition-consequence structure — "The customers don't agree on the product requirements, so we'll only be able to satisfy one of our major customers" — provides a better look at the big picture.

Figure 5.8   Risk documentation form.

ID: (Sequence number is fine.)
Description: (List each major risk facing the project. Describe each risk in the form "condition—consequence.")

Probability: (What's the likelihood of this risk becoming a problem?)	**Loss:** (What's the damage if the risk becomes a problem?)	**Exposure:** (Multiply Probability times Loss to estimate the risk exposure.)

First Indicator: (Describe the earliest indicator or trigger condition that might indicate that the risk is turning into a problem.)
Mitigation Approaches: (State one or more approaches to control, avoid, minimize, or otherwise mitigate the risk.)

Owner: (Assign each risk mitigation action to an individual for resolution.)	**Date Due:** (State a date by which the mitigation approach is to be implemented.)

The second row of cells in Figure 5.8 provides locations to capture the likelihood of a risk materializing into a problem (P), the loss you could incur as a result of that problem (L), and the overall risk exposure (E = P multiplied by L). Keep the high-exposure items at the top of your priority list. You can't address every risk item, so use this prioritization mechanism to learn where to focus your risk control energy. Set goals for determining when each risk item has been satisfactorily controlled. For some items, your mitigation strategies may focus on reducing the probability, while the approach for other items may emphasize reducing the loss.

The cell labeled Mitigation Approaches lets you identify the actions you will take to control the risk item. With any luck, you can use some of your mitigation approaches to address multiple risk factors. For example, one group I've worked with identified several risks related to failures of components of their web delivery infrastructure (servers, firewall, e-mail interface, and so on). To mitigate those risks, the group decided to implement an automated monitoring system that could check the status of the servers and communication functions periodically and alert them to any failures.

As with other project management activities, you need to get into a rhythm of periodic monitoring. You may wish to appoint a "risk czar" for the project, who is responsible for staying on top of the things that could go wrong, just as the project manager is staying on top of the activities leading to project completion. One company I know assigned an individual to such a role, and dubbed him "Eeyore," after the Winnie the Pooh character who always bemoaned how bad things could become.

Keep the top 10 risks highly visible, and track the effectiveness of your mitigation approaches regularly. As the initial list of top priority items gradually gets beaten into submission, new items may float up into the top 10. Don't kid yourself into concluding that you've controlled a risk simply because you've completed the selected mitigation action. Controlling a risk may mean you have to change your mitigation strategy if you conclude it's ineffective.

You can drop the risk off your threat detection radar when you determine your mitigation approaches have indeed reduced the loss exposure from that item to an acceptable level.

Common Schedule Risks

- Feature creep
- Requirements or developer gold-plating
- Shortchanged quality
- Overly optimistic schedules
- Inadequate design
- Silver-bullet syndrome
- Research-oriented development
- Weak personnel selection
- Contractor failure
- Friction between developers and customers

This information is adapted from Steve McConnell's *Rapid Development* (Microsoft Press, 1996).

Risk Management Can Be Your Friend

The skillful project manager uses risk management to raise awareness of conditions that could cause a project to go down the tubes. Consider a project that begins with an unclear product vision and a lack of customer involvement. The astute project manager will spot this situation as posing potential risks, and will document it in the risk management plan. Early in the project's life, the impact of this situation may not be too severe. However, if time continues to pass and the lack of product vision and customer involvement are not improved, the risk exposure will steadily rise.

By reviewing the risk management plan periodically, the project manager can adjust the probability and impact of these risks. Those that don't get controlled can be brought to the attention of senior managers or other stakeholders who are in a position to either stimulate corrective actions or make a conscious business decision to proceed in spite of the risks. Keep your eyes open and make informed decisions, even if you can't control every adverse condition facing your project.

Elements of Risk Management

You can address risks at any of the following levels:

1. Schedule risks. Best addressed at levels four and five, otherwise you'll lose the schedule battle.
2. Crisis management. Address risks only after they have become problematic.
3. Fix on failure. Detect and react to risks quickly, but only after they've occurred.
4. Risk mitigation. Plan ahead to provide resources to cover risks if they occur, but do nothing to eliminate them first.

5. Prevention. Implement and execute a plan as part of the project to identify risks and prevent them from becoming problems.

6. Elimination of root causes. Identify and eliminate factors that enable risks to occur.

This information is adapted from Steve McConnell's *Rapid Development* (Microsoft Press, 1996).

Learning from the Past

While you can't predict exactly which project threats might come to pass, you can do a better job of learning from previous experiences to avoid the same pain and suffering on future projects. As you begin to implement risk management approaches, keep records of your actions and results for future reference. Try these suggestions:

- Record the results of even informal risk assessments to capture the thinking of the project participants.
- Document the mitigation strategies attempted for each risk you chose to confront, noting which approaches worked well and which didn't pay off.
- Conduct post-project reviews (including retrospectives) to identify the unanticipated problems that arose. Would you have seen them coming through a better risk management approach, or would you have been blindsided in any case? Do you think these same problems might occur on other projects? If so, add them to your growing checklist of potential risk factors to think about for the next project.

Anything you can do to improve your ability to avoid or minimize problems on future projects will improve your company's business success. Risk management can also reduce the chaos, frustration, and constant firefighting that reduces the quality of life in so many organizations. The risks are out there. Find them before they find you.

5.5.6 "Estimating Internet Development"

by William Roetzheim

Online projects encompass market strategy and business process analysis, throwing traditional metrics out of whack.

Today, the fate of many companies depends on accurately predicting time-to-market for mission-critical Internet systems. The rate of change in the Internet and intranet development world is unprecedented, and estimating tools, methodologies and metrics are hard pressed to keep up. While traditional estimating, especially for client/server development, has significant applicability to Internet projects, there are some crucial differences. The largest issue is the fact that Internet software development projects now encompass activities that were previously done prior to the software project, namely market-strategy formulation and business-process analysis. The melding of these two domains with the software development project creates a hybrid that requires adjustments to the parametric models, the life-cycle templates and the standard deliverable lists.

What's more, because of competitive pressures, Internet and intranet business sites are often given high priority, with senior management imposing unrealistic deadlines. Completion time may be based on nothing more formal than wishful thinking, yet, because the site deployment is integrated with many other aspects of business operations, schedule slips can have dire financial consequences. Before we explore how to plan in Internet time, however, let's review the basics of heuristic and parametric estimating.

Estimating Fundamentals

Regardless of the type of application, the obvious first step in project planning is estimating the program volume or size, using any one of a variety of heuristic and parametric approaches. Heuristic, or trial and error methods, include top-down estimating — in which a cost for the design is predetermined and then allocated among the project activities — and bottom-up estimating — in which a list of all tasks is generated and a gut feeling or Delphi approach used to determine costs of each. Parametric approaches, on the other hand, use measures that can be approximated early in the project life and that should have a correlation with the final effort (see Figure 5.9). There are two main parameters for estimating volume: lines of code (LOC) and function points. The first is applicable to systems where the effort is driven by the complexity of the algorithms, or where the application has little or no user interface. Examples include embedded systems, real-time systems and many complex scientific applications. Function points are determined by looking at the number of reports, user data entry screens, tables in the database and so on. Function points are appropriate for data-driven systems (MIS, accounting and most client/server systems). Two less common parametric methods are GUI metrics (using menus, dialogs, windows and so on) and object metrics (using the number of objects).

Once the volume has been estimated, a series of formulas is used to convert the volume into initial, unadjusted estimates of effort and duration (see "Apply coefficients" in Figure 5.9).

Figure 5.9 Overview of a parametric estimating cycle.

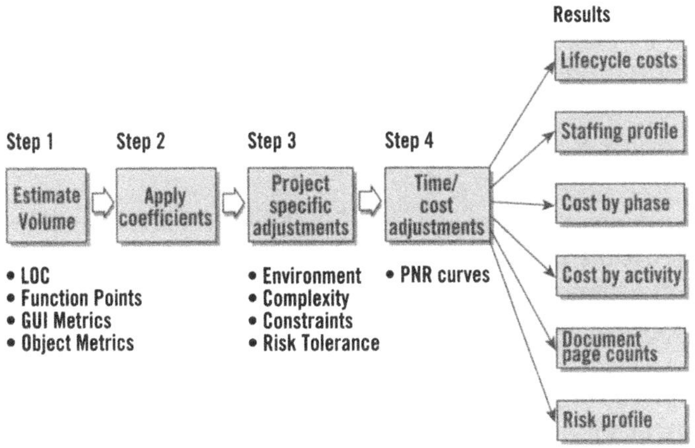

Because each project is unique in one or more ways, project-specific adjustments must then be made. These include the development environment, complexity, managerial constraints and risk tolerance of the organization. The result will be an adjusted estimate of the effort and an optimal delivery schedule.

Putnam-Norden-Rayleigh (PNR) curves (see Figure 5.10) show that a project schedule can be compressed or expanded within a range of 75 to 200 percent. Compression of greater than 75 percent is not feasible without changing some of the fundamental characteristics of the project (volume, environment and so on). Expansion beyond 200 percent results in increased costs and is not normally practical. If there are constraints on the project that require the job be completed at other than the optimal schedule, then the final schedule can be adjusted, but the costs must also be adjusted by a corresponding amount.

Figure 5.10   The project time/cost tradeoff (derived from PNR curves).

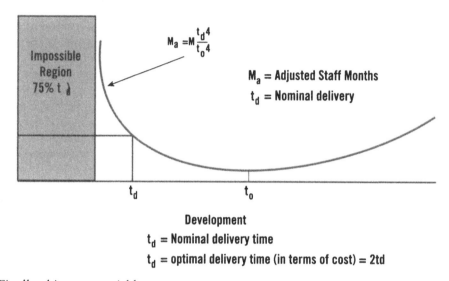

Finally, this process yields:

- Development and life-cycle costs.
- Optimum staffing curves by labor category.
- Cost broken down by phase and time.
- Cost allocation to individual activities in the work breakdown structure.
- A deliverable document list with descriptions and projected page counts.
- A risk profile for the project.

This approach to project estimating is valid for both traditional and Internet development jobs. However, the standard parametric models need some work before they can be put in place for Internet development projects. Let's look at how Internet development differs from traditional development.

Internet vs. Traditional Development

Most aspects of Internet projects are analogous to client/server applications. The systems tend to involve significant effort for database design, business rules and business objects, and interfaces to other systems. Parametric models based on function points are typically the most appropriate, with a life cycle containing client/server development tasks working the best. Examples of suitable life cycles include client/server (large), client server (small), IBM's

SMSLC, James Martin's Information Engineering, Marotz X-Force Development and Oracle's CASE*Method.

However, my focus is not on the similarities between Internet and traditional client/server applications, but rather the differences. In the "old fashioned" world of software development — that is, only a few years ago — there were three distinct types of projects and people (project stakeholders) supporting those projects. First, marketing and advertising consultants — be they internal or external — worked on product positioning and branding, corporate and product look-and-feel, and market demographics and segmentation. Second, business consultants worked on supply chain management, inventory management and business process reengineering. Finally, software consultants worked on database design, business object and rule design, and system architecture.

Surprisingly, there was little integration between these three groups, due perhaps to vastly different cultures. The marketing and advertising team is likely to be extremely creative, artistic, given to intuitive conclusions and driven by individual effort. The business consultants are likely to be comprised of formal and careful observers, strong communicators, and more facilitators than doers. Finally, the software engineers are typically anticorporate establishment, very structured and process-oriented, and analytical to a fault. However, the gestation time for new corporate initiatives used to be three to five years, which allowed the results of the work from all three groups to slowly blend at the senior management level and result in software projects that were more or less compatible with the work of the marketing and business consultants.

Now, enter the Internet world. Successful Internet projects require the immediate and complete integration of all three areas. The marketing and advertising types define the company's product positioning and branding, look and feel, and customer segmentation approach — all of which are executed in the new set of Internet applications. The Internet pushes the company to integrate its supply chains, front office systems and back office systems to be able to automatically promote, sell and deliver the products to the customer. And the software developers must build systems that are every bit as complicated as traditional client/server systems, but must also be the execution arm of the marketing and the business consultants.

To make matters much worse, the timeframe for delivering the new systems has been compressed from several years to — typically — several months. Today, these three groups must work closely together for the system to successfully meet the corporate objectives. The impact of this transition, of course, affects software estimation. We'll look at the impact from the perspective of the basic parametric models, the system life cycle templates, the environmental adjustments and the documentation requirements.

Parametric Internet Models

Selecting a suitable parametric estimating metric is one of the most fundamental decisions in cost estimating. For some development environments and tools, there may exist a metric tailored to that type of development. For example, Domino Points is a parametric model optimized for Lotus Domino development. Other Internet-compatible models are coming on the scene, including:

- *Internet Points* — an extension of Function Points to explicitly support Web site development.
- *Use-Case Points* — an early model of size using use cases in accordance with Rational's Unified Process model.

- *Class-Method Points* — a design model of size using Rational's Unified Process model.

If you're not currently using an Internet-specific cost model for your online projects, in virtually all cases the appropriate choice is function points. When using function points for an Internet development project, the following interpretations are appropriate:

- Inputs are each input screen or form (for example, CGI or Java), each maintenance screen, and if you use a tab notebook metaphor anywhere, each tab.

- Outputs are each static Web page, each dynamic web page script (for example, ASP, ISAPI, or other DHTML script), and each report (whether Web based or administrative in nature).

- Tables are each logical table in the database plus, if you are using XML to store data in a file, each XML object (or collection of XML attributes).

- Interfaces retain their definition as logical files (for example, unique record formats) into our out-of-the-system boundaries.

- Queries are each externally published or use a message-oriented interface. A typical example is DCOM or COM external references.

Table 5.6   A selection of software life-cycle templates.

Life cycle	Description
Domino — Web	Lotus Domino development for a Web application.
Domino — Workflow	Lotus Domino development for a workflow application.
IBM Net Commerce	IBM Net Commerce system development and setup.
Internet — E-Commerce 1	Internet development of an e-commerce solution using a rapid application development approach.
Internet — E-Commerce 2	Internet development of an e-commerce solution with extensive back-end work. Based on Roger Forunier's *A Methodology for Client/Server and Web Application Development* (Yourdon Press, 1998).
Internet — Web	Internet development of a Web site where most of the effort is on the front-end and user interface issues.
Rational Unified Process	Development in accordance with the Rational Unified Process.
Telecom — E-Commerce	Telecom e-commerce project, including automated interfaces to partners such as carriers and local exchange carriers.
Telecom — Web	Telecom customer service and liaison Web development.

Size and Effort

Because many estimating models use source lines of code (SLOC) as an input, it is useful to convert the function points to equivalent SLOC. This process is called backfiring and simply entails using a language-specific multiplier that maps function points (or another volume measure) to equivalent SLOC required to implement each function point. This process is neither new nor unique to Internet development. What is new, however, is the fact that the pace of change in the Internet development world means that new languages, and variations of existing languages, are released at a staggering pace. Because the backfiring multiples require some completed projects to compute the correct values, the list of available languages may not include the languages, or language versions, that you are using to build the system.

Finally, the estimating models use one or more efficiency factors to convert from the sizing metric (for example, function points or SLOC) to effort. Many models use two factors, a linear multiplier I'll call M(A) and a nonlinear adjustment factor I'll call M(B). M(A) defines the base efficiency of developing the function points, SLOC, Domino Points, Internet Points or whatever. M(B) determines how this efficiency changes with varying project sizes.

We have already seen that Internet development projects include activities that are not part of a traditional software development life cycle, such as market and business process analysis. Because of this, the cost per unit of size must obviously be greater for an Internet development project than for a traditional software project. This means the M(A) must be larger. In addition, we might speculate that the behavior of the efficiency function with project size (M(B)) might be different when comparing an Internet project to a traditional client server project. Unfortunately, insufficient data exists to either prove or disprove this hypothesis; for the time being, the M(B) factor is typically the same as, or similar to, the M(B) that would be used for a client/server project.

Environmental Adjustments

Costing models adjust for project and company specific environmental factors such as the complexity of the application, the skill of the development team, how widely dispersed the development team is, and so on. Published models use between eight and 20 factors (plus multiple subfactors) with COCOMO (short for the Constructive Cost Model, proposed by Barry W. Boehm in *Software Engineering Economics*, Prentice Hall, 1981) at the high end of the spectrum, standard function points at the low end and models such as MK-II function points in the middle. Internet-specific environmental factors to be considered include graphics and multimedia, legacy integration, site security, text content, tool selection, transaction loads and Web strategy.

Therefore, the parametric estimating models for Internet development vary from traditional client server development models in the following ways:

- New metrics may be introduced, such as Domino Points or Internet Points.
- New languages need to be supported, such as XML and Cold Fusion.
- The M(A) costing factor needs to be adjusted to compensate for the additional effort involved in nontraditional software development activities that are now intrinsic to the software project.
- New Internet specific environmental adjustments apply.

In the end, we have an estimate of effort and time. The next step in the estimating life cycle is to allocate that effort to individual work breakdown structure (WBS) activities in a project life cycle (or template).

Internet System Life Cycles

Project life cycles are work breakdown structure (WBS) templates that contain lists of tasks and cost allocations. They are used for project management, work package preparation, project tracking and so on. Traditionally, e-commerce sites use client/server development life-cycle templates and Web sites use simple waterfall life-cycle templates. However, the body of knowledge about e-commerce and Web development has grown to the point where, with more than 36 different industry standard life cycles now available (see Table 5.6), templates specific to these projects are recognized and accepted.

Documentation Requirements

Just as a life cycle defines a project template work breakdown structure, a standard defines a standard set of deliverable documentation (called artifacts by the Rational Unified Process) that will be created. As with life cycles, there are at least 43 software development standards.

Internet development projects are often critical to the market and business strategy of an organization, so effective estimates of completion dates, plus reduced changes of failure, are vital to the financial health of an organization. Luckily, state of the art cost estimating models now support development in this complex and rapidly changing environment.

5.5.7 "Web Time Software Development"

by Dave Thomas, edited by Larry Constantine

---

*Developing software in web time can be an immense undertaking.
Here's some advice on making it a smooth process
for everyone involved.*

The pace of software product development has changed dramatically over the past few years. The excitement of new technologies and new business opportunities has created a gold rush culture that pulls products from development at unprecedented rates. Gone are the days of one major new release every 12 to 18 months. To be competitive, companies must deliver products in web time. In web time, a year is only three or four months long.

Unfortunately, even the most advanced RAD processes and tools don't stand up to the sustained demand of quarterly releases. Such schedules are achievable only by the superhuman effort of talented software professionals. Coaching a web time development team is an immense undertaking. How many sports host a play-off schedule every quarter?

First-Time Processes

Process improvement literature aside, there is no repeatable software process for first-time software. For many new products, substantial portions of the product and its development are being created for the first time. Web time development is characterized by bootstrapping all the way to the release date. In my experience, it consists of guerrilla programming in a hostile environment using unproven tools, processes, and technology — both software and hardware. It is not unusual for a web time project to employ new people and to have a new product requirement.

Finally, to make development even more challenging, the product must typically have a new user interface that is flashy and completely original. While initial evaluations are often only skin deep, these interfaces are nonetheless important to the marketing department, especially when a web page announcement for a nonexistent product can yield a following of 100,000 potential customers.

Assessing Technical Risks

In a fast-changing World Wide Web environment, it's imperative to identify the technical risks and manage them properly. Be forewarned that you may need to educate your management and customers about these risks. In a macho, web time world, your concern may be perceived simply as an excuse for not delivering on an aggressive schedule. Failure due to instability, reliability, or performance in a number of areas — for example, platforms, middleware, tools, libraries, and integration with other products — can lead to missed promises with career-limiting results.

It's prudent to maintain a list of unknowns and monitor them carefully; seldom are they solved without cost. The inconsistency, poor quality, and lack of documentation for existing platforms require extensive behavioral experiments and intimate systems knowledge that rival the worst days of mainframe systems programming. Don't be surprised if you must acquire and retain the highly specialized expertise for one or more of the platforms or tools you are using.

Death by Quality

With the nature of web time, coupled with first-time software, goals cannot be achieved with cumbersome, top-down, process-oriented techniques, regardless of the amount of quality injected into the product. These techniques, such as total quality management, draw their strength from manufacturing standardized widgets, where repeatability is essential for continuous improvement.

In web time, quality must be instilled from the bottom up using talented, highly motivated, collaborative development teams — popularly referred to as high-performance teams. Software quality circles of small, four- to eight-person teams are key to surviving web time development. The teams must touch everyone who is involved with the product: customer, architect, product coordinator, development executive, developer, tester, technical writer, and installer.

Just-In-Time Software

Many manufacturers have responded to crises in their industries by following the just-in-time (JIT) process. At OTI, we borrow extensively from JIT, making the necessary adjustments to continuously adapt the process to the people rather than the people to the process. We have been evolving the process for the past 10 years with customers whose applications range from embedded systems to banking.

Our process is a proven, customer-driven approach for building products using high-performance teams. It uses an interactive, model-driven approach to software architecture, design, and development. The emphasis is on predictable delivery that stresses timely results over features.

JIT software is a disciplined engineering approach based on the well-documented lessons of three decades of software development. It is architecture-driven because an architectural process defines the major subsystems and their interfaces. The process identifies components and assigns each one an owner who is responsible for its design, implementation, testing, performance, and overall quality. Component ownership is a critical success factor, although the ownership may and does change during the 24-hour-a-day, seven-day-a-week cycles that are essential to develop, integrate, and test the product.

Our process is based on a manufacturing approach to software construction in that we build a product from existing components. The product evolves through a series of application models (or prototypes, as we constructively used to call them in engineering). Often the cycle begins with a display prototype to wow everyone with the nifty user interface. Then comes an interaction prototype, which hooks activities to the user interface to give it a particular feel. This is followed by a simulation prototype, which contains realistic models (stubs) for the product's essential behavior. Finally, the simulation prototype evolves into a design prototype for the final product.

Guidelines, Specifications, Reviews

There is a tendency to view JIT software as a polite form of hacking as opposed to the disciplined process of iterative refinement. My company's process replaces the always out-of-date software process descriptions of total quality management with lightweight, mutually accepted guidelines and reviews that support high-performance teams. We use the term "guideline" to reflect that guidelines are documents that record team consensus rather than standards imposed by management or the quality assurance department. Such guidelines are essential for communication and quality and need to be maintained on an intranet. Each team develops its own guidelines, which are based on its experience in previous projects and the challenges it faces. We also pair every team with an architect, who serves as both an advocate and a critic for the team.

Our primary specification tools are use cases, hypertext-based APIs, and interface descriptions. We describe class hierarchies with protocol-based interface specification and frameworks with design patterns.

It is impossible to perform in-depth reviews for all aspects of a project, so we use reviews as a method of risk reduction. Reviews can be initiated internally or externally on the basis of a perceived risk, such as complexity, vague requirements, or a developer's unknown skills or performance. We use automated quality assurance tools to identify code that may need further review. We review the test and release plans. Finally, beta-test customers review the product before we approve it for release.

Component and Assembly Life Cycles

Object or component technology is a necessary but insufficient condition for success in web time. To achieve high-quality software in three months, you must use components that were constructed and tested in previous time boxes. The current time box provides sufficient time only to assemble the components, update the interface, and repair major defects.

The critical success factor is recognizing that components and the product (application) have separate life cycles. Components must go through a life cycle from simulation to prototype, alpha testing, beta testing, and production.

Our process is environment-based in that all developers continuously sit in a build of the product and have instant access to all changes that are made in the development library. This access lets them incrementally roll their changes forward or backward with fine control. The process is supported by IBM's VisualAge, a proven environment for team development and configuration management. We use VisualAge to support not only our product development, but also our tools and other processes such as testing and problem reporting.

Managing by Time Boxes

Time is both your friend and your enemy in web time software development. We manage our development using three- to four-month time boxes, so the development time for a time box is between three and six weeks. Because instability from a new tool or platform can easily consume a time box's entire development time, management's most important decision is the decision of what to leave out.

With so little time to prepare or manage detailed development plans, we opt for a process that uses software developers and what I call "playing coaches." Each developer has a time box and an associated deliverable for which he or she is responsible. Customers prioritize features based on the trade-offs that the engineering teams identify. As a result, choosing features is not a developer's arbitrary decision. The resulting project schedule is the sum of the component time boxes plus the system test time box.

Test Once, Test Everywhere

The promise of Java is "write once, run everywhere (WORA)." But for those of us who have actually lived in the multi-platform, multi-geography world, we know that this really means "test once, test everywhere."

Lurking beneath the nirvana of portable APIs is the reality of trench warfare. Trying to solve insidious, undocumented problems transforms a straightforward programming assignment into an unbounded exploration into the interactions of operating systems, middleware, and tools. What is a new and exciting API for the press often is a minefield of undocumented bugs and features for developers.

Most organizations simply do not provide the infrastructure that is needed to support multi-platform and multi-culture development. Producing portable software requires a substantial investment in test configurations, frameworks, and tools. For various commercial reasons, from two to three releases of each operating system must be supported. In addition, it is well known that computer platforms, like pollinating trees, must be purchased in pairs so developers can understand and test them adequately.

We have developed processes and tools for international software development so that we can support both European and Asian releases with a single release cycle. A critical success factor for web time global software is access to engineering experts in each geographical area, which is itself sufficient justification for hiring students from around the globe.

Absorbing Standards

We live with a plethora of standards, which can be official (such as X-Window System, Motif, and POSIX) or de facto (such as Microsoft's ActiveX and JavaSoft's Java). Even a modest effort in tooling can provide the means to quickly absorb industry standards and reduce the error-prone manual implementation of a standard.

Our automated tools support a process to migrate standards definitions into standards-compliant object frameworks and associated test suites. The tools let us rapidly adopt a new standard for competitive advantage. Equally important, they let us track the standard and help us develop testing frameworks and platform regression testing.

Release Engineering

One of the most dramatic effects of web time on our software processes has been the increased importance and technical complexity of our release engineering process. In most traditional development processes, release engineering is performed by the development team as an end-of-cycle activity or by an often disconnected release engineering team as an independent activity.

Unlike the total quality management world of controlled releases, web time development often consists of almost continuous release. We have joked that we should hook our development library to our web site so we don't waste time baselining each release.

Release engineering is so important that we assign critical development management and architects to the activity. We've developed special tools to build releases for multiple platforms as well as supply incremental, corrective service updates.

Team leaders' war room meetings prioritize and ensure that all appropriate fixes (and only the appropriate fixes) are applied to the release. While we attempted to follow a traditional freeze-fix-release approach, the instability caused by so much concurrent development has required a disciplined sequence of freeze-thaw-freeze. Two senior developers must approve any post-freeze fix.

To reduce feature creep and compress the time to integrate and manufacture, you need deep technical knowledge to focus on the important problems. Further, both release architects and development executives must actively block feature creep, which invades through both the front and back doors as the marketing team and the developers feel compelled to include extra fixes and enhancements.

Peopleware

Unfortunately, web time development is possible only if you have high-performance teams. The obvious corollary is that there is little room for the average, steady worker or a manager who needs orderly plans and working hours. As a result, we demand increasingly more of experienced software developers. It isn't unusual for developers to work around the clock for extended periods. While the demands may be exciting for a new hire, more than ever before they can take their toll on the relationships and families of critical, seasoned developers.

Team members must understand the strengths and limitations of each other. They must measure each other by their ability to cope with and balance their personal and professional commitments rather than by the absolute hours they may work. Team members must watch out for each other, watch for signs of depression and excess stress.

Management must understand the effect on spouses and children and offer families a helping hand through special events and time out. At our company, we encourage children and dogs to visit the office so everyone can spend more time together during a hectic release schedule.

As we push the limits of web time software development into what resembles an extreme sport, we come close to the human limits. As development managers, however, we are far from the experience of Olympic coaches and their cadre of sports psychologists. Only by allowing rest time for our teams can we help developers sustain their efforts over the long term. For those who develop software in web time, we must seriously consider nine-month work years.

5.5.8 "Managing Outsourced Projects"

by Steve McConnell, edited by Larry Constantine

Learn these six basic steps to effective outsourcing and your efforts can be virtually problem-free.

Some of the greatest software project risks are associated with projects outsourced to internal or external vendors. How can you manage outsourced projects without losing control?

Consider this scenario, based on a true story. Giga-Corp.'s customers were clamoring for the second version of its popular software product, but it didn't have sufficient technical staff to work on the upgrade, and the manager in charge of the project was already overwhelmed. Giga-Corp. decided to outsource.

It created a 50-page requirements specification and sent a request for proposal to a dozen vendors. The request for proposal didn't include a budget, but did indicate a strong preference for a fixed price bid with delivery within five months. After holding a meeting to answer vendors' questions, Giga-Corp. received a handful of bids, all for time and materials, ranging from $300,000 to $1,200,000. The high bid came from Fledgling Inc., a startup founded by a former Giga-Corp. programmer who had worked on the first version of the software. Giga-Corp. threw it out, along with the low bid, on the assumption that those were outliers and probably the result of misunderstanding the project's technical requirements. It selected middle-of-the-pack Mega-Staff and then negotiated the estimate down to $350,000.

Heading South

The first undeniable hint of trouble came when Mega-Staff's head recruiter called Fledgling Inc.'s president on a Thursday afternoon, sounding desperate. "We need three C++ programmers by Monday," he said. "Giga-Corp. accepted our bid, but their senior engineer said our Visual Basic approach wasn't technically feasible and we don't have enough C++ programmers to support the project." Fledgling Inc.'s president said he couldn't find three C++ programmers on such short notice, and wished Mega-Staff's recruiter good luck.

Somehow, when Monday rolled around, Mega-Staff showed up at Giga-Corp. headquarters with three new C++ programmers. The Mega-Staff team was enthusiastic and appeared to make good progress at first, but by the five-month mark, the team was nowhere close to completion. Mega-Staff promised delivery within eight months, which was acceptable to Giga-Corp. because that coincided with the beginning of its next sales year. Eight months into the project the end was still far from reach, and Giga-Corp. demanded cost concessions for late delivery.

Fourteen months into the project, after more schedule slips and many painful meetings with Giga-Corp., Mega-Staff delivered about 25% of the planned functionality — at a cost of approximately $1,500,000. Considering the functionality underrun, Mega-Staff had essentially overrun its proposed cost by 1,400% and its proposed schedule by 1,000%.

Battening Down the Hatches

With software outsourcing on the rise, the Giga-Corp. and Mega-Staff disaster is being replayed by software outsourcing buyers and vendors throughout the world. These problems could have been avoided if the participants had understood the basics of the software outsourcing life cycle.

Effective outsourcing consists of six steps: specifying the requirements in detail, making an explicit decision to outsource, obtaining resources, selecting a vendor, writing the contract, and building an in-house management team that will monitor the project to completion. Super large projects — DoD style — will need more steps, but most business software projects will benefit from learning the specifics of these basic ones.

Step One: Specify Requirements in Detail

Because the requirements document often becomes part of the legal contract between the buyer and vendor, outsourced projects need high-quality requirements. If requirements aren't clearly specified in the beginning, the project can later become a battleground on which all possible ramifications of the buyer's and seller's different interests are contested. As the case study illustrates, this is a battle that both parties will lose.

If the organization doing the buying doesn't have the in-house resources to create detailed requirements, it should use a two-phase acquisition. During the first phase, the buyer outsources development of the requirements. During the second phase, the buyer puts those requirements out for bid and then outsources the development of the software. In general, if an organization doesn't have the technical or managerial resources in-house to perform the necessary tasks for effective outsourcing, it should have a disinterested third party perform them.

Detailed requirements also form the basis of the buyer's effort and schedule estimates. A buyer who doesn't create detailed estimates or have them prepared by a disinterested third party won't know whether a $300,000 or a $1,200,000 bid is more realistic.

Step Two: Make an Explicit Decision to Outsource

The decision to outsource invariably involves weighing both costs and benefits. Because outsourcing will increase a project's risk, you should make the decision explicitly, with full awareness of potential problems. Don't choose to outsource because it seems to be the easiest option — it rarely seems easy by the time the project is complete.

Step Three: Obtain Resources

The buyer must acquire sufficient resources to complete the project. These resources include budgets for the proposal phase, vendor payment, and management of the project on the buyer's side.

Step Four: Select a Vendor to Develop the System

Choose a software outsourcing vendor carefully. The vendor selection process usually entails creating a request for proposal, distributing it to potential vendors, and requesting that they submit proposals. The request for proposal should contain at least the following materials:

Software requirements specification (SRS). These are the detailed requirements that were developed in the first step to effective outsourcing.

Statement of work (SOW). The statement of work is a document containing the management requirements for the project. It specifies your demands regarding managing changes to requirements, what kinds of tools and development environments will be used, technical methods and methodologies, software documentation, engineering data (design notes, diagrams, scaffolding code, build scripts, and so on), backup procedures, source code control, quality assurance procedures, and the project's management organization — especially the interaction between your organization and the vendor.

Documentation list. The request for proposal should specify the documents you want to have developed in conjunction with the software. You should include design documents, source code documentation, operating manuals, status reports, and any other documents you want developed.

Cost and schedule estimates. Include the cost and schedule estimates you prepared during requirements development. Giga-Corp. didn't know how much its project should cost or how long it should take, which led it to hire a vendor incapable of performing the work. If you don't publish your cost and schedule estimates in the request for proposal, the proposal process degenerates into a game in which each vendor tries to guess your budget.

The divergence among vendor estimates to Giga-Corp. suggests that either the technical requirements were not developed well enough or that some or all of the vendors were poor estimators. The fact that the highest bid was submitted by the company most familiar with the software was a clear warning. If Giga-Corp. didn't want to choose the highest bidder, it would have been wise to switch to a two-phase acquisition approach, using one of the vendors to develop the requirements more fully, creating new cost and schedule estimates, and then putting the project out to bid again.

Evaluation criteria. Tell the vendors what criteria you will use to evaluate their proposals. Typical criteria include project management capability, general technical capability, technical design approach, technical methodologies, technical documentation, engineering data management, requirements management approach, configuration management approach, and quality assurance approach.

Competent vendors will explain in their proposals how they plan to meet each of the evaluation criteria. By publishing evaluation criteria, you make it easier for yourself to make side-by-side comparisons among proposals, and you significantly improve the quality of information that vendors provide.

Proposal preparation guidelines. Describe how you want the proposals to be organized and formatted. Include descriptions and page limits for each section, margin sizes, contents of running headers and footers, and font size. This might seem overly picky, but specifying a standard proposal format makes the job of proposal evaluation easier.

Be sure to create a sample proposal so that you know the page count limits are reasonable. On larger projects, buyers distribute their proposal guidelines for review by vendors before they distribute the official request for proposal. This gives vendors a chance to comment on the page count, evaluation criteria, and other issues. It also helps improve the quality of information the buyer ultimately obtains through the proposal process.

Picking the Winner

If you did your homework when you created the request for proposal, proposal evaluation will be nearly painless. Create a decision matrix based on the evaluation criteria described in your request for proposal, and score each proposal accordingly. Be prepared to follow up with questions to cover missing or insufficient proposal responses.

You might eliminate some vendors because of low scores overall or in specific categories. For example, you might eliminate any vendor that scores less than 10 out of 15 points in project management capability. If Mega-Staff scored only seven points in project management capability, it would have been eliminated. By publishing explicit evaluation criteria and scoring proposals based on them, Giga-Corp. could have known that Mega-Staff was critically weak in project management. Thus, Mega-Staff's eventual project overrun would have been predictable — and avoidable — at proposal time.

As the Giga-Corp. case illustrates, the lowest bid and the final project price are often not related. The goal of awarding a contract is not to choose the lowest bid but to choose the vendor that will provide the best value — do the best job for the least money as judged by systematic evaluations of the proposals.

You can also use the evaluation criteria for negotiating changes to the winning vendor's proposal. For instance, Giga-Corp. could have negotiated changes to requirements management and quality assurance approaches if those were the weak points in the winning bid.

Step Five: Write the Contract

Your warm feelings about your choice of vendor will quickly evaporate if the project begins to unravel. Be sure your contract spells out the details of management requirements, technical requirements, warranties, patents and other intellectual property issues, contract termination, payment, and any other important issues. Don't try to draw up the contract yourself. You can lose in court if your contract doesn't mean what you think it means, so spend the money to have it reviewed by legal counsel before you sign it.

Step Six: In-House Management and Monitoring

The magnitude of Mega-Staff's overrun suggests that Giga-Corp. essentially didn't monitor or control the project after the contract was awarded. Giga-Corp.'s managers might have voluntarily accepted a 100% or 200% cost overrun, but they would have to be trying to put themselves out of business to knowingly accept an overrun of 1,400%.

The most common mistake I see in managing outsourced software development is that no one on the buyer side manages the outsourced development project at all. While outsourcing can indeed reduce the amount of management needed, it increases the degree of management sophistication needed. The problems involved with managing a project across the hall are magnified when you have to manage a project across town or across the globe.

Because you can't monitor progress of an outsourced project just by walking around and talking to project team members, project tracking indicators must be more formal. In your request for proposal, you should specify what management controls you want the vendor to provide so you can track the project's progress. Such controls might include: weekly status reports; weekly updates to the list of technical risks; weekly defect statistics; weekly statistics on the amount of code added, modified, and deleted; weekly reports on the number of modules planned, number coded, and number that have passed their code reviews; and any of many other indicators of project health.

On small projects, someone from the buyer's side must be put in charge of monitoring the vendor's progress and sounding an alarm when problems arise. On large projects, several people from the buyer's side might be needed.

Smooth Sailing

These steps might seem like a lot of work for an outsourced project. As the case study illustrates, however, the price of not doing your homework can be budget and schedule overruns of 1,000% or more. The time you spend laying the groundwork for a successful project and then steering the project to the desired conclusion is time well spent. Just ask any project manager who has been drawn by the siren song of easy software to the rocky shores of unreliable vendors, cost and schedule overruns, functionality underruns, and failed projects.

5.5.9 "Selecting the Best Vendor"

by Neil Whitten

Outside vendors can be a boon to your software project — or a burden — depending on how you handle the relationship.

Software development organizations are increasingly looking to outside companies or "vendors" to help them complete their projects faster, cheaper, and with improved quality. A software development manager can encounter many problems when working with vendors. (In this article, a vendor is an outside, independent company hired to perform some type of software development work, such as developing, testing, and supporting one or more components of a product, or perhaps even the whole product.)

Managers often believe that because vendors operate as independent entities, management can exercise little influence over the direction and quality of the vendor's work. Usually this is true because of poor planning of the vendor-selection process. Selecting the vendor best suited to your needs is the first step to creating a positive vendor relationship. Equally important is a good vendor contract. A well-thought-out contract will spell out clearly what you want and what you expect from the vendor and enable you to maintain the necessary control of the project and the vendor's work.

Creating a Good Vendor Contract

Figure 5.11 shows the steps of the vendor contract process. Step 1 is the Request for Proposal (RFP) — you may use a different term for the RFP at your company. The RFP addresses two primary topics: the Statement of Work (SOW)and the instructions to bidders (vendors). The SOW describes the work the vendor will perform, when that work is due, how the work will be measured for acceptance, and the working relationship between the vendor and your organization. The SOW becomes a major part of the contract between the vendor and your organization.

The instructions to bidders provides vendors with the information they need to create and submit a bid proposal, such as format, submission media, key dates, and contact people. It also contains information on the categories that each proposal will be evaluated against and the process they can expect your company to follow in selecting the best candidate.

In step 2, the candidate vendors create their proposals based on the information provided in the RFP and submit them to your organization. In step 3, the vendor's proposals are evaluated, the vendor is selected and notified, and the unsuccessful bidders are informed of the strengths and weaknesses of their proposals. In step 4, you write the contract, negotiate its terms, and obtain an agreement. After the agreement, the contract is executed according to its terms and conditions.

Figure 5.11   The vendor contract process.

1 Software group creates RFP and distributes it to vendors

2 Vendors create proposals and deliver to software group

3 Software group selects vendor

4 Software group and vendor negotiate, sign, and execute contract

Defining the Vendor's Work

A direct relationship exists between the completeness of your RFP and the completeness of the bid proposals you'll receive. Therefore, you must provide candidates with sufficient information about your project. The SOW must fully describe the work the vendor will perform. The better you define the job, the better estimate a vendor can submit. If you provide an incomplete description of the work you want performed, you can expect months of problems with the vendor as you attempt to define the job at hand.

For example, it's helpful to provide complete product specifications in the RFP, such as the functions to be provided along with their proposed user interfaces, and the requirements for topics such as hardware and software to be supported, system resources and performance, installation, product and programming standards, compatibility and migration, and security.

It is typical to provide a start date for the contract work and ask the bidder to provide the best delivery date possible. Here, define your quality level for the product (for example, two defects per every thousand lines of code delivered into system test), and ask the vendor to provide a realistic schedule that will yield a product at this quality level.

Don't forget to clearly define what you mean by "end of project" and "end of effort" if these two have different meanings. For example, you might define "end of project" as the point when the product has completed testing and is packaged and ready for delivery to the first customer. "End of effort" may mean the point at which the vendor is no longer required to maintain the code. This could be the end of the project or two years after the code has been in use by customers. Again, you want vendors to understand exactly what they're committing to so you receive a more realistic bid.

You should also ask vendors to identify any trade-offs they feel they must make between quality, schedule, and cost. Some creative ideas may be offered to save time and money. Beware, sacrificing quality for schedules and costs almost never pays.

Figure 5.12 Evaluating categories.

Category	Vendor ABC	Vendor DEF	Vendor GHI
Technical	2.43	2.54	2.10
Project Management	1.49	2.27	1.93
Quality	1.33	2.33	1.75
Business	2.14	2.73	2.18
Average score	1.85	2.47	1.99

Share the News

Don't blindside the people in your organization with the RFP. Before releasing the RFP to vendors, show it to the project groups that will be affected by its terms and conditions and obtain their support. These groups might be development, publications, testing, finance, and quality assurance. Here's a good rule of thumb for the RFP and the subsequent contract: If it is important, document it. If it is not written down, then it probably will not be performed, and you'll have difficulty binding the vendor to do it.

Selecting a Vendor

Do not rush through the vendor selection process. Selecting a vendor can be a painstaking chore, but the time you invest to select the best candidate can become one of the best investments you will make. The time and money you spend creating and reviewing the RFP, sending it to vendors, evaluating their proposals, and so on is insignificant compared to the amount of time you spend working with this vendor and the impact the vendor will have on your product.

Here are some actions to keep in mind as you search for a vendor:

- Always seek bids from at least three vendors. Just as your competition keeps you on your toes to provide high-quality, low-cost products and services, so too are vendors challenged by competing vendors. Take advantage of this competitive environment and seek bids from at least three vendors (whenever possible) for every one vendor you are looking to select.

- Make sure vendor candidates are qualified. The vendors you choose from must meet your technical and security requirements, confidential disclosure agreements, general business terms and conditions, quality standards, and deadlines — and any other requirement you consider essential to the project.

- Prepare a complete and well-thought-out proposal. This point was introduced previously, but is so important it bears repeating. Vendors are responding to the proposal you provide them. If your proposal is not complete, is not clear, or otherwise inadequate, expect the vendor responses to be the same. Provide vendors with the information they need to provide you with a thorough bid. At the very beginning of the relationship with vendors, set a professional example in letting them know what you expect from them. You will find that this investment will pay itself back many times over while working with the selected vendor.

- Define an objective bid proposal process. Before the RFP document is completed and given to the candidate vendors, define the process you will follow to evaluate your candidates. Identify the categories you'll examine most closely and make sure the competing vendors understand these categories (see Figure 5.12 and Figure 5.13). This planning will help you develop a well-thought-out RFP that identifies the major areas you want the vendors to address.

- Don't assume anything when evaluating a proposal. If you're uncertain about anything, ask the vendor.

The Bid Process

You can evaluate the proposals many ways. One option is to involve a selected group of people from the organization, led by a chairperson. Each member of the group reviews the proposals — either the entire proposal or a specific category in which they have expertise such as technical or project management — and determines a score for each category featured in the RFP.

Figure 5.13 shows an example of scoring for the category of project managment. Three vendors (ABC, DEF, and GHI) are evaluated. The scoring approach, shown in Figure 5.14, uses the values 3, 2, 1, and 0, where 3 is the highest score. The numbers assigned to the subtopics (the bulleted items in Figure 5.13) are averaged to determine the category scores (Project Scheduling, Tracking, and Reporting Methodology, Project Organization Structure, and so on). The topic scores are also averaged to determine the category score for project management. Finally, the category scores are averaged to determine the overall score for each vendor. These overall scores are compared to select the best vendor candidate. In the example shown in Figure 5.12, vendor DEF has achieved the highest score.

The lowest-dollar bidder is not necessarily the best vendor to select. To avoid any "cost-effective" bias, don't let committee members see the cost proposals during the evaluation process. This way, they'll identify the best candidate regardless of cost. Factor in the cost later to help make the final decision. After all the vendors have been notified of your decision, work with the selected vendor to resolve any problems and finalize the contract.

A Changing World

Rapidly changing technologies, global competition, shortage of critical skills, and economic considerations are but a few reasons why vendors are becoming increasingly popular in software development. Thoughtful and deliberate planning of the vendor-selection process is essential to ensure that you select the best vendor for your project and make the most of the benefits they can bring to your organization.

Figure 5.13 Evaluating the project management category.

Project Management Evaluation Topics	Vendor ABC	Vendor DEF	Vendor GHI
Project Scheduling, Tracking, and Reporting Methodology	**1.75**	**2.25**	**2.00**
• Clearly defined process, roles, responsibilities, tools	2	3	2
• Frequent and regular tracking meetings	2	2	2
• Dependency management	1	2	2
• Planned communication with company buying services	2	2	2
Project Organization Structure	**2.00**	**2.50**	**2.00**
• Supports software development process; built-in checks and balances	2	2	2
• Clear definition of roles and authority	2	3	2
Problem Management Approach	**2.00**	**2.00**	**2.00**
• Defined and responsive process	2	2	2
Staffing and Skills	**1.00**	**2.00**	**2.00**
• Availability of headcount, skills, leaders across required diciplines	1	2	2
Related Experience	**1.00**	**2.50**	**2.00**
• Seasoned development; experience on similar products	1	3	2
• Indicators of learning from past experience	---	2	2
Software Development Process	**1.67**	**2.67**	**2.00**
• Clearly defined activities, roles, and responsibilities	2	3	2
• Comprehensive	2	3	2
• Effective checks and balances, entry and exit conditions	1	2	2
Facilities, Equipment, and Tools	**2.00**	**2.00**	**1.50**
• Productive environment	2	2	2
• Security	2	2	1
Average score for project management	**1.49**	**2.27**	**1.93**

Figure 5.14 Scoring criteria.

Score	Meaning	Definition
3	Exceptional	Exceeds performance or capability; high probability of success; no significant weaknesses
2	Acceptable	Satisfies evaluation criteria; good probability of success; any weaknesses can be easily corrected
1	Marginal	Currently fails to satisfy the evaluation criteria; low probability of success; significant deficiencies, but correctable
0	Unacceptable	Significant deficiencies that are difficult to correct

5.5.10 "A Software Metrics Primer"

by Karl Wiegers

Combining metrics with process improvement can help you define the areas where you need to improve the quality of your team's applications.

My friend Nicole is a quality manager at Motorola, a company widely known for its software process improvement and measurement success. Once she said, "Our latest code inspection only found two major defects, but we expected to find five, so we're trying to figure out what's going on." Few organizations have enough insight into their software development and project management efforts to make such precise statements. Nicole's organization spent years making its processes repeatable and defined, measuring critical aspects of its work, and using those measurements to set up well-managed projects that develop high-quality products.

Software measurement is a challenging but essential component of a healthy and capable software engineering culture. In this article, I will describe some basic software measurement principles and suggest some metrics that can help you understand and improve the way your organization operates. Planning and implementing your measurement activities can take significant effort, but the payoff will come over time.

Why Measure Software?

Software projects are notorious for running over schedule and budget, and still ending up with quality problems. Software measurement lets you quantify your schedule, work effort, product size, project status, and quality performance. If you don't measure your current performance and use the data to improve your future work estimates, those estimates will just be guesses. Because today's current data becomes tomorrow's historical data, it's never too late to start recording key information about your project.

You can't track project status meaningfully unless you know the actual effort and time spent on each task compared to your plans. You can't sensibly decide whether your product is stable enough to ship unless you're tracking the rates at which your team is finding and fixing defects. You can't quantify how well your new development processes are working without some measure of your current performance and a baseline to compare against. Metrics help you better control your software projects, and learn more about the way your organization works.

What to Measure

You can measure many aspects of your software products, projects, and processes. The trick is to select a small and balanced set of metrics that will help your organization track progress toward its goals. Goal-Question-Metric (GQM) is an excellent technique for selecting appropriate metrics to meet your needs. With GQM, you begin by selecting a few project or organizational goals. State the goals to be as quantitative and measurable as you can. They might include things like:

- Reduce maintenance costs by 50% within one year.
- Improve schedule estimation accuracy to within 10% of actual.
- Reduce system testing time by three weeks on the next project.
- Reduce the time to close a defect by 40% within three months.

For each goal, think of questions you would have to answer to see if you are reaching that goal. If your goal was "reduce maintenance costs by 50% within one year," these might be some appropriate questions:

- How much do we spend on maintenance each month?
- What fraction of our maintenance costs do we spend on each application we support?
- How much money do we spend on adaptive (adapting to a changed environment), perfective (adding enhancements), and corrective (fixing defects) maintenance?

Finally, identify metrics that will let you answer each question. Some of these will be simple data items you can count directly, such as the total budget spent on maintenance. Other metrics you will compute from two or more data items. To answer the last question listed previously, you must know the hours spent on each of the three maintenance activity types and the total maintenance cost over a period of time.

Notice I expressed several goals in terms of a percentage change from the current level. The first step of a metrics program is to establish a current baseline, so you can track progress against it and toward your goals. I prefer relative improvement goals ("reduce maintenance by 50%") to absolute goals ("reduce maintenance to 10% of total effort"). You can probably reduce maintenance to 10% of total effort within a year if you are currently at 20%, but not if you spend 80% of your effort on maintenance today.

Your balanced metrics set should eventually include items relating to product size, product quality, process quality, work effort, project status, and customer satisfaction. Table 5.7 suggests metrics that individual developers, project teams, and development organizations should consider collecting. You would track most of these over time. For example, your routine project tracking activities should monitor the percentage of requirements implemented and tested, the number of open and closed defects, and so on. You can't start with all of these, but I recommend including at least the following measurements early in your metrics program:

- *Product size:* count lines of code, function points, object classes, number of requirements, or GUI elements

- *Estimated and actual duration (calendar time) and effort (labor hours):* track for individual tasks, project milestones, and overall product development

- *Work effort distribution:* record the time spent in development activities (project management, requirements specification, design, coding, and testing) and maintenance activities (adaptive, perfective, and corrective)

- *Defects:* count the number of defects found by testing and by customers, and the defects' type, severity, and status (open or closed).

Table 5.7   Appropriate metrics for software developers, teams, and organizations.

Group	Appropriate Metrics
Developers	Work Effort Distribution Estimated vs. Actual Task Duration and Effort Code Covered by Unit Testing Number of Defects Found by Unit Testing Code and Design Complexity
Project Teams	Product Size Work Effort Distribution Requirements Status (Number Approved, Implemented, and Verified) Percentage of Test Cases Passed Estimated vs. Actual Duration Between Major Milestones Estimated vs. Actual Staffing Levels Number of Defects Found by Integration and System Testing Number of Defects Found by Inspections Defect Status Requirements Stability Number of Tasks Planned and Completed
Development Teams	Released Defect Levels Product Development Cycle Time Schedule and Effort Estimating Accuracy Reuse Effectiveness Planned and Actual Cost

Creating a Measurement Culture

Fear is often a software practitioner's first reaction to a new metrics program. People are afraid the data will be used against them, that it will take too much time to collect and analyze the data, or that the team will fixate on getting the numbers right rather than building good software. Creating a software measurement culture and overcoming such resistance will take diligent, consistent steering by managers who are committed to metrics and sensitive to these concerns.

To help your team overcome the fear, you must educate them about the metrics program. Tell them why measurement is important and how you intend to use the data. Make it clear that you will never use metrics data either to punish or reward individuals (and then make sure that you don't). A competent software manager does not need individual metrics to distinguish the effective team contributors from the slackers.

Respect the privacy of the data. It is harder to abuse the data if managers don't know who the data came from, Classify each data item you collect into one of these three privacy levels:

- *Individual:* only the individual who collected the data about his or her own work knows it is his or her data, although it may be pooled with data from other individuals to provide an overall project profile.
- *Project team:* data is private to the members of the project team, although it may be pooled with data from other projects to provide an overall organizational profile.
- *Organization:* all members of the organization can share the data.

As an example, if you're collecting work effort distribution data, the number of hours each individual spends working on every development or maintenance phase activity in a week is private to that individual. The total distribution of hours from all team members is private to the project team, and the distribution across all projects is public to everyone in the organization. View and present the data items that are private to individuals only in the aggregate or as averages over the group.

Make It a Habit

Software measurement doesn't need to be time-consuming. Commercial tools are available for measuring code size and complexity in many programming languages. Activities like daily time tracking are more of a habit than a burden. Commercial problem tracking tools facilitate counting defects and tracking their status, but this requires the discipline to report all identified defects and to manage them with the tool. Develop simple tracking forms, scripts, and web-based reporting tools to reduce the overhead of collecting and reporting the data. Use spreadsheets and charts to track and report on the accumulated data at regular intervals.

Tips for Metrics Success

Despite the challenges, many software organizations routinely measure aspects of their work. If you wish to join this club, keep the following tips in mind.

Start small. Because developing your measurement culture and infrastructure will take time, use GQM to first select a basic set of initial metrics. Once your team becomes used to the idea of measurement and you have established momentum, you can introduce new metrics that will give you the additional information you need to manage your projects and organization effectively.

A risk with any metrics activity is dysfunctional measurement, in which participants alter their behavior to optimize something that is being measured, rather than focusing on the real organizational goal. For example, if you are measuring productivity but not quality, expect some developers to change their coding style to expand the volume of material they produce, or to code quickly without regard for bugs. I can write code very fast if it doesn't actually have to run correctly. The balanced set of measurements helps prevent dysfunctional behavior by monitoring the group's performance in several complementary aspects of their work that lead to project success. Never attempt to use metrics to motivate performance.

Explain why. Be prepared to explain to a skeptical team why you wish to measure the items you choose. They have the right to understand your motivations and why you think the data will be valuable. Use the data that is collected, rather than letting it rot in the dark recesses of a write-only database.

Share the data. Your team will be more motivated to participate in measurement activities if you inform them about how you've used the data. Share summaries and trends with the team at regular intervals and get them to help you understand what the data is telling you. Let them know whenever you've been able to use their data to answer a question, make a prediction, or assist your management efforts.

Define data items and procedures. It is more difficult and time-consuming to precisely define the data items and metrics than you might think. However, if you don't pin these definitions down, participants may interpret and apply them in different ways. Define what you mean by a "line of code," spell out which activities go into the various work effort categories, and agree on what a "defect" is. Write clear, succinct procedures for collecting and reporting the measures you select.

Understand trends. Trends in the data over time are more significant than individual data points. Some trends may be subject to multiple interpretations. Did the number of defects per function point found during system testing decrease because the latest round of testing was ineffective, or did fewer defects slip past development into the testing process? Make sure you understand what the data is telling you, but don't rationalize your observations away.

See the Big Picture

Link your measurement program to your organizational goals and process improvement program. Use the process improvement initiative to choose which improvements to focus on. Use the metrics program to track progress toward your improvement goals, and use your recognition program to motivate and reward desired behaviors. Measurement alone will not improve your team's performance, but it will provide information that will help you focus and evaluate your software process improvement efforts.

5.5.11 "Metrics: 10 Traps to Avoid"

by Karl Wiegers

By being aware of these 10 common risks, you can chart a course toward successful measurement of your organization's software development activities.

As software development gradually evolves from art to engineering, more and more developers appreciate the importance of measuring the work they do. While software metrics can help you understand and improve your work, implementing a metrics program is a challenge. Both the technical and the human aspects of software measurement are difficult to manage.

According to metrics guru Howard Rubin, up to 80% of software metrics initiatives fail. This article identifies 10 traps that can sabotage the unsuspecting metrics practitioner. Several symptoms of each trap are described, along with some possible solutions. By being aware of these common risks, you can chart a course toward successful measurement of your organization's software development activities.

Trap One: Lack of Management Commitment

Symptoms: As with most major improvement initiatives, management commitment is essential for a metrics effort to succeed. The most obvious symptom that commitment is lacking is when your management actively opposes measurement. More frequently, management claims to support measurement, and effort is devoted to designing a program, but practitioners do not collect data because management hasn't explicitly required it of them.

Another clue that managers aren't fully committed is when they charter a metrics program and planning team, but then do not assist with deploying the program into practice. Managers who are not committed to software measurement will not use the available data to help them do a better job, and they won't share the data trends with the developers in the group.

Solutions: Educate managers so that they understand the value of software measurement. A good resource is *Practical Software Metrics for Project Management and Process Improvement* by Robert Grady (PTR Prentice-Hall, 1992). Tie the metrics program to the managers' business goals so they see that good data provides the only way to tell if the organization is becoming more effective. You need management's input to help design the metrics program to ensure it will meet their needs.

If you cannot obtain commitment from senior management, turn your attention to the project and individual practitioner level. Developers and project teams can use many valuable metrics to understand and improve their work, so focus your energy on those people who are willing to try. As with any improvement initiative, grass roots efforts can be effective locally, and you can use positive results to encourage a broader level of awareness and commitment.

Trap Two: Measuring Too Much, Too Soon

Symptoms: You can measure hundreds of aspects of software products, projects, and processes. It is easy to select too many different data items to be collected when beginning a metrics program. You may not be able to properly analyze the data as fast as it pours in, so the excess data is simply wasted. Those who receive your summary charts and reports may be overwhelmed by the volume and they may tune out.

Until a measurement mindset is established in the organization, expect resistance to the concept of measuring software as well as the time required to collect, report, and interpret the data. Developers who are new to software measurement may not believe that you really need all the data items you are requesting. A long list of metrics can scare off some of the managers and practitioners whose participation you need for the program to succeed.

Solutions: Begin growing your measurement culture by selecting a fairly small, balanced set of software metrics. By balanced, I mean that you are measuring several complementary aspects of your work, such as quality, complexity, and schedule. As your team members learn what the metrics program is all about, and how the data will (and will not) be used, you can gradually expand the suite of metrics being collected. Start simple and build on your successes.

The participants in the metrics program must understand why the requested metrics are valuable before they'll want to do their part. For each metric you propose, ask, "What can we do differently if we have this data?" If you can't come up with an answer, perhaps you don't need to measure that particular item right away. Once the participants are in the habit of using the data they collect to help them understand their work and make better decisions, you can expand the program.

Trap Three: Measuring Too Little, Too Late

Symptoms: Some programs start by collecting just one or two data items, which may not provide enough useful information to let people understand what's going on and make better decisions. This can lead stakeholders to conclude that the metrics effort is not worthwhile, so they terminate it prematurely.

Another obstacle to getting adequate and timely data is the resistance many software people exhibit toward metrics. Participants who are more comfortable working undercover may drag their feet on measurement. They may report only a few of the requested data items, report only data points that make them look good, or turn in their data long after it is due. The result is that managers and project leaders don't get the data they need in a timely way.

A metrics program has the potential to do damage if too few dimensions of your work are being measured. People are tempted to change their behavior in reaction to what is being measured, which can have unfortunate and unanticipated side effects. For example, if you are measuring productivity but not quality, some people may change their programming style to generate more lines of code and therefore look more productive. I can write code very fast if the quality is not important.

Solutions: As with trap two, the balanced set of metrics is essential. Measure several aspects of your product size, work effort, project status, product quality, or customer satisfaction. You don't need to start with all of these at once. Instead, select a small suite of key measures that will help you understand your group's work better, and begin collecting them right away. Since software metrics are usually a lagging indicator of what is going on, the later you start, the farther off-track your project might stray before you realize it. Avoid choosing metrics that might tempt program participants to optimize one aspect of their performance at the expense of others. Make participation in the metrics program a job expectation.

Trap Four: Measuring the Wrong Things

Symptoms: Do the data items being collected clearly relate to the key success strategies for your business? Are your managers obtaining the timely information they need to do a better job of managing their projects and people? Can you tell from the data whether the process changes you have made are working? If not, it's time to reevaluate your metrics suite. Another symptom of this trap is that inappropriate surrogate measures are being used. One example is attempting to measure actual project work effort using an accounting system that insists upon 40 labor hours per week per employee.

Solutions: Select measures that will help you steer your process improvement activities by showing whether or not process changes are having the desired effect. For example, if you're taking steps to reduce the backlog of change requests, measure the total number of requests submitted, the number open each week, and the average days each request is open. To evaluate your quality control processes, count the number of defects found in each test and inspection stage, as well as the defects reported by customers.

Make sure you know who the audience is for the metrics data, and make sure the metrics you collect will accurately answer their questions. As you design the program, leverage from what individuals or project teams are already measuring. The goal/question/metric paradigm works well for selecting those metrics that will let you answer specific questions associated with organizational or project goals. A good example of applying the goal/question/metric paradigm is found in "A Practical View of Software Measurement and Implementation Experiences Within Motorola" by Michael K. Daskalantonakis (*IEEE Transactions on Software Engineering*, Nov. 1992).

Trap Five: Imprecise Metrics Definitions

Symptoms: Every practitioner interprets vague or ambiguous metric definitions differently. One person counts an unnecessary software feature as a defect, while someone else does not. Time spent fixing a bug found by testing is classified as a test effort by one person, a coding effort by another, and rework by a third. Trends in metrics tracked over time may show erratic behavior because individuals are not measuring, reporting, or charting their results in the same way.

You'll have a clue that your metrics definitions are inadequate if participants are frequently puzzled about what exactly they are being expected to measure. If you keep getting questions like, "Do I count unpaid overtime in the total work effort?" you may be falling into this trap.

Solutions: A complete and consistent set of definitions for the things you are trying to measure is essential if you wish to combine data from several individuals or projects. For example, the definition of a line of code is by no means standard even for a single programming language. Standardize on a single tool for collecting metrics based on source code. Automate the measurement process where possible, and use standard calibration files to make sure all participants have configured their tools correctly.

Those designing the metrics program must create a precise definition for each data item being collected, as well as for other metrics computed from combinations of these data items. This is much harder than you might suspect. Plan to spend considerable time agreeing on definitions, documenting them as clearly as possible, and writing procedures to assist practitioners with collecting the data items easily and accurately.

Trap Six: Using Metrics Data to Evaluate Individuals

Symptoms: The kiss of death for a software metrics initiative is to use the data as input into an individual's performance evaluation. Using metrics data for either reward or punishment, such as rank-ordering programmers based on their lines of code generated per day, is completely inappropriate. When people think the numbers they report might be used against them, they'll either stop reporting numbers at all, or only report numbers that make them look good. Fear of the consequences of reporting honest data is a root of many metrics program failures.

Solutions: Management must make it clear that the purposes of the metrics program are to understand how software is being built, to permit informed decisions to be made, and to assess the impact of process changes on the software work. The purpose is not to evaluate individual team members. Control the scope of visibility of different kinds of data; if individual names are not attached to the numbers, individual evaluations cannot be made. However, it is appropriate to include the activity of collecting and using accurate (and expected) data in an individual's performance evaluation.

Certain metrics should be private to the individual; an example is the number of defects found by unit testing or code review. Others should remain private to a project team, such as the percentage of requirements successfully tested and the number of requirements changes. Some metrics should have management visibility beyond the project, including actual vs. estimated schedule and budget, and the number of customer-reported defects. The best results are achieved when individuals use their private metrics data to judge and correct themselves, thereby improving their personal software process.

Trap Seven: Using Metrics to Motivate Rather Than to Understand

Symptoms: When managers attempt to use a measurement program as a tool for motivating desired behaviors, they may reward people or projects based on performance with regard to just one or two metrics. Public tracking charts may be pointed out as showing desirable or undesirable results. This can cause practitioners who are using the charts to understand what's up with their software to hide their data, thereby avoiding the risk of public management scrutiny. Managers may focus on getting the numbers where they want them to be instead of really hearing what the data is telling them. As with trap three, the behavioral changes stimulated by motivational measurement may not be the ones you really want.

Solutions: Metrics data is intrinsically neither virtuous nor evil; it's simply informative. Using metrics to motivate rather than to learn has the potential of leading to dysfunctional behavior, in which the results obtained are not consistent with the goals intended by the motivator. Metrics dysfunction can include inappropriately optimizing one software dimension at the expense of others, or reporting fabricated data to tell managers what they want to hear.

Stress to the participants that the data must be accurate if you are to understand the current reality and take appropriate actions. Use the data to understand discrepancies between your quality and productivity goals and the current reality so you can improve your processes accordingly. Use the process improvement program as the tool to drive desired behaviors, and use the metrics program to see if you are getting the results you want. If you still choose to use measurement to help motivate desired behaviors, be very careful.

Trap Eight: Collecting Unused Data

Symptoms: The members of your organization may diligently collect the data and report it as requested, yet they never see evidence that it is being used. People may grumble that they spend precious time measuring their work, but they don't have any idea why this is necessary. Often, the required data is submitted to some collection center, which stores it in a write-only database. Your management doesn't seem to care whether data is reported or not, similar to trap one.

Solutions: Software engineering should be a data-driven profession, but it can't be if the available data disappears from sight. Project leaders and upper management must close the loop and share results with the team. They must relate the benefits of having the data available, and describe how the information helped managers make good decisions and take appropriate actions. Selected public metrics trends must be made visible to all stakeholders, so they can share in the successes and choose how to address the shortcomings.

Today's current data becomes tomorrow's historical data, and future projects can use previous results to improve estimating capabilities. Make sure your team members know how the data is being used. Give them access to the public metrics repository so they can view it — and use it — themselves.

Trap Nine: Lack of Communication and Training

Symptoms: You may be falling into this trap if the participants in the metrics program don't understand what is expected of them, or if you hear a lot of opposition to the program. Fear of measurement is a classic sign that the objectives and intent of the program need to be better communicated. If people do not understand the measurements and have not been trained in how to perform them, they won't collect reliable data at the right times.

Solutions: Create a short training class to provide some basic background on software metrics, describe your program, and clarify each participant's role. Explain the individual data items being collected and how they will be used. Defuse the fear of measurement by stressing that individuals will not be evaluated on the basis of any software metrics data. Develop a handbook and web site with detailed definitions and procedures for each of the requested data items. Top-down communication from management should stress the need for data-driven decision-making as well as the need to create a measurement-friendly culture.

Trap Ten: Misinterpreting Metrics Data

Symptoms: A metric trend that jumps in an undesirable direction can stimulate a knee-jerk response to take some action to get the metric back on track. Conversely, metrics trends that warn of serious problems may be ignored by those who don't want to hear bad news. For example, if your defect densities increase despite quality improvement efforts, you might conclude the "improvements" are doing more harm than good, and be tempted to revert to old ways of working. In reality, improved testing might well find a larger fraction of the present defects — this is good!

Solutions: Monitor the trends that key metrics exhibit over time, and don't overreact to single data points. Make sure you understand the error bars around each of your measures so you can tell whether a trend reversal is significant. If you can figure out why (say, your post-release bug correction effort has increased in two successive calendar quarters), you can decide whether corrective action is required. Allow time for the data you're collecting to settle into trends, and make sure you understand what the data is telling you before you change your course of action.

Many of us struggle with how to implement a sensible metrics program that gives us the information we need to manage our projects and organizations more effectively. Staying alert to the 10 risks described in this article can increase the chance of successfully implementing a software metrics initiative in your organization.

5.5.12 "Don't Fence Me In"

by Warren Keuffel

In many ways, the field of object orientation remains uncharted territory. A few pioneers have begun to explore the possibilities of metrics use in this relatively new domain.

Over the past few months, we've examined a wide variety of metrics ranging from effort measurements, such as lines of code, to predictor measurements, such as function points. One active area of interest we haven't touched on yet is the question of how we apply these tools to the object-oriented environment.

Object orientation provides some interesting challenges for the individual attempting to apply engineering principles — which find their fundamental core in measurement — to this environment. It is clear that predictor metrics, most notably function points, can map equally well to the object-oriented as well as the non-object-oriented development environments because they focus on the functionality delivered to the user.

Code attributes such as inheritance, polymorphism, and encapsulation are irrelevant to the function point counter whose interest, you will recall, is focused on the boundaries between the system and the external sources of input and output (whether users or other computer systems) as well as the internal files and other entities that must be employed in the system. For example, it is irrelevant whether the developer employed code reuse except insofar as such reuse positively affects the cost to deliver a particular function point.

But predictor metrics tell us nothing about a system's internal complexity. If a system is built with classes that presumably employ object-oriented tools such as encapsulation, polymorphism, and inheritance, the traditional procedural tools such as McCabe's cyclomatic complexity and Halstead's software science only scratch the surface and reveal little of the complexity buried in the class structures. Hence, it is not surprising that measurement researchers and practitioners have started to turn their attentions toward developing and identifying measurements suitable for languages such as Java, Smalltalk, and C++.

What is surprising to me, however, is the paucity of good sources on the subject. I'm going to briefly discuss two references you might wish to consider if you are interested in pursuing this subject further. But I must caution you that the field of object-oriented metrics appears to resemble nothing more than a frontier on which pioneers are busily staking out their claims to immortality. As such, no metrics sheriff exists to protect you from unsubstantiated claims. This frontier analogy has other dimensions as well. Consider that object-oriented programmers often wear their "cowboy" reputations proudly; their disdain for conventional (read: restrictive) software engineering is still prevalent — although fading, as object orientation becomes absorbed into the mainstream of software development.

Chidamber and Kemerer Blaze the Trail

The most cited reference I've found is the paper "A Metrics Suite for Object-Oriented Design" (*IEEE Transactions on Software Engineering*, June 1994), by S.R. Chidamber and C.F. Kemerer of Massachusetts Institute of Technology. (Chidamber and Kemerer also presented an earlier version of this work at an OOPSLA conference.) Chidamber and Kemerer's paper focuses on coupling and cohesion as two important complexity attributes of a given class structure, and they define five measurements that provide insight into these attributes:

- *Weighted methods per class.* Chi-damber and Kemerer suggest that we give more complex methods greater weight, but they fail to provide weighting criteria.

- *Depth of inheritance tree.* How deep is the ancestry of the class?

- *Number of children.* The number of subclasses immediately subordinate to the class you are measuring.

- *Coupling between object classes.* As Larry Constantine pointed out in *Structured Design* (Yourdon Press, 1975) more than 20 years ago, high coupling between modules decreases the benefits of encapsulation; the same is true of objects.

- *Response for a class.* The total number of responses possible from all methods of a class.

Chidamber and Kemerer cheerfully admit that their work is based on theory, not experience, and invite the reader to validate and expand upon their work. For the most part, the measurements they suggest lend themselves to automated data collection. Unfortunately, they present no cause-and-effect relationship data to tie these metrics into attributes of greater practical interest, such as maintainability or reusability.

Despite wide interest in the Chi-damber and Kemerer metrics, some prominent methodologists in the field have serious reservations about them. For example, Brian Henderson-Sellers and Larry Constantine, both at the University of Technology in Sydney, are preparing a paper that details some problems they have found in the mathematical formulae used to calculate object cohesion metrics. So the interest is certainly there in the academic community, but what practical guidelines exist for the practicing professional?

Lorenz and Kidd: Shotgunning Wagonmasters

In contrast to the theoretical approach offered by Chidamber and Kemerer, *Object-Oriented Software Metrics* by Mark Lorenz and Jeff Kidd (Prentice-Hall, 1994) offers the practitioner a veritable smorgasbord of both project and design metrics. Lorenz and Kidd worked in IBM's Object-Oriented Technology Center before striking out on their own. Lorenz and Kidd offer so many measurements (they provide eight project metrics and 30 design metrics) that you hardly know where to begin. In some regards, this represents the shotgun approach to measurement — in which you load up as many metrics as will fit in the gun and blast away, hoping that at least one of the measurements will hit something interesting. Because Lorenz's company offers a measurement tool that captures the recommended data, I can't help but wonder if the book is an elaborate sales aid for the product.

Despite these reservations, as of today, Lorenz and Kidd have written what is arguably the most useful guide to applying metrics to the object-ori ented environment. Because it is based on the writers' real-world experience with the metrics, it provides an immediate reference unavailable elsewhere. The downside is that it is based primarily on anecdotal evidence and the writers' own project database, so it does not reflect a truly validated suite of measurement tools.

Tom DeMarco, Metrics Gunflinger

As I write this, I've just returned from the fifth annual Applications of Software Measurement (ASM) conference, where Tom DeMarco delivered the keynote address. It is probably stretching the "Old West" analogy to call DeMarco a gunflinger, because he's a "down easter" from

Maine, and as he gets older he adopts more and more curmudgeonly behaviors. That's not to say that curmudgeons can't be right occasionally, and in this case DeMarco's talk, "Mad about Measurement," certainly gave the conference attendees something to think about. As you might expect, the title was deliciously ambiguous: were we to be angry about measurement or ecstatic about it? As soon as I saw the title I knew the answer would be the former, having witnessed several of DeMarco's blasts at unneeded documentation (such as ISO-9000) or infringements on First Amendment rights.

And I was right. Weaving various skeins of measurement misconduct anecdotes into a web, DeMarco led up to his central thesis that measurement unchecked could well be leading us back to the excesses of management by objectives, or MBO. Robert McNamara, the U.S. secretary of defense who architected the Vietnam War, was a firm believer in MBO; his primary metric was the body count — a metric that was directly responsible for excesses such as the My Lai massacre. Closer to home, A.J. Demming, a guru of the quality movement, has castigated reliance on MBO as destroying the process ownership so necessary for delivering high-quality products. DeMarco drove this point home with a devastating comparison between IBM, which has enthusiastically accepted ISO-9000, and Microsoft, which has essentially thumbed its nose at the ISO standards.

Measurement Homesteaders

DeMarco's talk, however, was pure icing on a substantial cake of solid papers and presentations from and for practicing metrics professionals. If you're serious about measurement, be sure to consider attending this annual Software Quality Engineering-sponsored event. Because Barbara Kitchenham became ill and was unable to deliver her scheduled talk on simplifying function points, SQE president Bill Hetzel stepped into the breach with some interesting data on metrics utilization gathered from conference participants over the previous five years.

I've received more mail from people asking me to help their organizations set up metrics programs than on any other topic in recent memory. If you feel you're on the trailing end of the metrics curve, take heart in some of Hetzel's findings. Out of 1,200 participants in the five-year survey, only 2% were satisfied with their organization's measurement program; 50% described their company's program as poor (and these were organizations motivated enough to send their people to a metrics conference!).

If you're debating whether to set up a separate metrics group, consider that 34% of those organizations with separate groups described their groups' effectiveness as good or excellent, compared to 8% of those organizations that did not maintain separate measurement groups. But even those organizations that captured a variety of measurements found the measurements most valued by the metrics professionals were among the least used, indicating that establishing a metrics group is no guarantee an organization will successfully incorporate the generated data into its process feedback loops.

Function Point Tribes

A few weeks prior to the ASM conference, the International Function Point Users Group (IFPUG) held its semi-annual conference in Salt Lake City, so I was easily able to attend. These are a nice bunch of people with many good things going for them, but, like many tribes, the group has developed an "us vs. them" mentality that ultimately may work against it. In all the sessions, the emphasis was on how to use function points, how to get started with a function point counting effort, or the like. Obviously, as this was a function point conference, the emphasis was correctly placed, but as someone pointed out once on CASEFORUM, if you can't clearly describe three ways to abuse a tool, you haven't mastered it. And this was what was most prominently missing from the IFPUG presentations: discussions of when *not* to use function points — and why.

Despite this carping, if you're interested in function points — and they are useful tools, probably the best tools we have for estimating effort — the IFPUG is a good tribe to join. If you're coming from a corporate environment, the IFPUG membership fee structure is great: $250 per year for an unlimited number of participants. If you work for yourself, the picture is less rosy: $250 per member. I lobbied with some of the IFPUG board members to adopt a membership strategy that will encourage more individual participation; we will see what response that generates.

Mail Outlaws

Meanwhile, back at the ranch, I've encountered a governmental process that's clearly out of control. Like many other people, I had the post office forward my mail to a mail agent — one of the many storefronts where you can send and receive UPS, Federal Express, and U.S. mail. I had the unfortunate luck to choose a mail agent that suddenly and without warning ceased business. No problem, I thought. I'll just have the friendly folks at the post office forward my mail to my home. Uh-oh. I quickly found out that when you establish a relationship with a mail agent you sign *away* your right to have your mail forwarded to another address — even if your relationship with the mail agent was involuntarily terminated. Despite an effort by my congresswoman's staffer on my behalf, the post office refused to budge, and persisted in bouncing all my mail back to sender. And the Post Office wonders why nobody likes their service.

If you're a vendor or publisher and have had your press releases or software or books returned, now you know why. If you need my current address, please send me e-mail. And remember, if you sign up with a mail agent, you're gambling that the agent will stay in business as long as you need it. It's time to ride off into the sunset after completing this column. Until next month, see you on the bitstream.

5.5.13 "Software Measurement: What's In It for Me?"

by Arlene Minkiewicz

Recently, while attending a conference, I overheard a number of software quality and process professionals lamenting the fact that the software measurement programs for which they were responsible for were doomed to fail. Discussion with these managers led me to believe that the problem wasn't poorly conceived or poorly executed measurement programs, but poorly accepted ones. The biggest obstacle to implementing a successful software measurement program is getting the developers — the ones in the trenches designing, coding, documenting and testing the software — to measure their work.

To many software developers, not only is there no obvious incentive to a measurement program, there is a disincentive of having their output scrutinized and possibly misunderstood. Who wants his boss to know that the guy in the next cubicle wrote more code than he did last week? If measurement programs are to work, software professionals must understand the personal and professional benefits of these programs.

Why the Resistance?

I have been in the software development business long enough to appreciate the value of a well-implemented software measurement program. This appreciation has grown as I have progressed through my career from programmer to analyst to manager of software development teams. Indeed, it would be fair to say that early in my career I viewed measurement programs with skepticism and maybe even a little contempt.

Here are some of the reasons why the typical software developer might resist attempts to institute a measurement program:

If you are measuring my output, then you may try to use this as an input to my performance evaluation (and thus your measurement will in some way impact my salary). Developers understand that measuring what they do requires a more detailed understanding of software and the development life cycle than is often expressed in the traditional measurements applied to their output.

Measuring my output takes time away from what you hired me to do: develop software. If developers are expected to spend all their time measuring how long it took them to develop last week's software, where will they find the time to develop this week's software? Software development is a real-time occupation — if I stop to measure, I will lose valuable momentum, or so the thinking goes.

What does input into the measurement program buy me? Why should I spend my time and put my reputation (translate: ego) on the line to help you develop a measurement program when the only results of this program I can see are more work, more frustration and possibly less pay for me?

With each of these objections, it's clear where the resistance lies. While most software measurement programs count on the cooperation of the software developers, many fail to address the concerns of these same people. If you develop software or work with developers, you know what I mean when I say software developers are often quite intense. They take tremendous pride in their ability to build something from nothing. They prefer to spend their working day in "The Zone," giving their full concentration to the current problem, oblivious to conversations, office politics, telephone calls and incoming e-mails. In general, these developers despise what they see as meaningless paperwork because it detracts from the main mission — creating software. If we, as the project and quality assurance managers, don't look at software measurement from the developer's perspective, we will never succeed.

Show Me the Money

So do software developers truly benefit from measurement programs? We all know the benefits of effective measurement to the software project, the software project manager and the organization that uses or sells the software. We often don't bother to think of the benefits these same programs might offer to the software developer. Rather than ignoring these benefits, we should identify and use them to sell the measurement programs to the developers. Measurement programs are intended to improve organizational processes in order to make higher quality software in less time with less effort. Certainly, the main benefit to the organization can be seen as a benefit to the developer as well, since the anticipated result will put more money in everybody's pockets. But there are also secondary benefits for the developer as well. With defined processes in place, they'll find themselves less subject to the side effects of poor planning. They are less likely to end up working on failures and more likely to work on projects with clearly defined goals and an established software development process — in other words, projects they enjoy.

Selling to the Measurers

What steps can you take to ensure that the software designers, developers, testers and others in your organization buy into your measurement program? Start by having clear goals for the measurement program and making sure that everyone on the team understands what these goals are and what the organization expects to gain. Then follow up. Make sure that there are periodic reviews of the measurement program to determine whether or not these goals are being met. Make the results of these reviews available in concise, easy-to-understand terms. Celebrate successes but don't hide failures. If goals aren't being met, show how the problems are being addressed and what changes are planned. No one wants to take time away from the important and enjoyable parts of his or her job to do paperwork that feeds a rubber-stamp process. Therefore, show how the measurement program is helping to meet specific organizational goals. Relate it to the bottom line wherever possible.

Next, involve everyone in planning the measurement program. Include the measurers in the decisions about what to measure and how to measure it. This benefits your program in two ways. First, the measurers will feel more ownership in the program if they have had real input from the beginning. Second, they will have valuable input into which measurements are easy to make and which ones will be impossible to measure consistently and accurately. You may find resistance to some of your measurements, but this, too, is information you should know up front. It may be necessary to rethink some of the measurement goals or to consider additional automation investments. You may find that you get suggestions for alternate measurements or measurement processes that make good sense for your program and cost substantially less in time to accomplish. Developers know what they know and know how best to get to it.

Support the measurement program from the top down. Software developers, like everyone else, are evaluated against specific goals. If measurement tasks aren't included in my goals but I am asked to include measurement as a deliverable, I will probably spend an inadequate amount of time working on measurement tasks. If, whenever there is a schedule crunch, my manager tells me to shelve measurement tasks and concentrate on the current schedule goal, the seriousness with which I approach measurement tasks will diminish. Measurers must see commitment from above. At every milestone, deliverables should include specific measurement data that are as important to the milestone as the requirements document, design document or executable. Everyone on the team should have specific goals that relate to measurement and that are considered along with other goals when performance is evaluated. In order to get the folks in the trenches to take measurement seriously, managers need to take it seriously. It can't be the first thing that gets pushed aside in a crunch.

This is much easier said than done. We often develop software against aggressive, inflexible schedule constraints. And when the business is riding on timely delivery, how important is it that we stop and report how much time we spent on each task and how many defects we delivered in the design? The answer depends on how long you expect your business to be in business and how many more times you want to be in similar schedule crunches. Of course, you shouldn't sacrifice your business to make your measurement program succeed. On the other hand, make sure that exceptions to the measurement rules are carefully considered and rarely occur.

Relate successful measurement to better working conditions. Having spent many years writing code myself, I have occasionally found myself in the office late at night trying to finish a module or track down a problem. Although sometimes I have only myself to blame, there have been instances when I find myself wishing my boss had done a better job of planning the project or selling his plans to his boss. I suspect I'm not the only software developer who has found herself rushing to catch up to a schedule that didn't make a whole lot of sense in the first place. Over time I have come to realize that I could have helped my boss plan projects better by supplying him with accurate software measurements against which future plans could be based. Not only has this helped to reduce schedule pressure, it has also resulted in my participation in calmer projects with much more predictable outcomes. It's amazing how much easier it is to estimate the future accurately with data from the past. It's equally amazing how much easier it is to negotiate a software plan with management if you have data (or previous estimating successes) to back it up.

This is not to say this will be an easy sell. You can't show improvements until you've done a fair amount of measurement, and you have to be committed to using the measurement in this way. If making more realistic schedule estimates is not one of the goals of your measurement program, don't present it as one. Some organizations use schedule pressure as a means of increasing productivity. (As an aside, I don't think this is a particularly effective long-term strategy and would recommend it only for those efforts already recognized as death-march projects.) If you belong to an organization like this, I wouldn't recommend trying to sell reduced schedule pressure as a benefit of the measurement program. If, however, reduced schedule pressure due to more realistic estimates is a benefit of your program, make sure you sell this aspect to your measurers — those who traditionally work late into the night as the end of a project draws near. Take care to manage expectations as well, since this is sure to be more of a long-term than a short-term goal, and you don't want to lose enthusiasm just as you are about to start making real progress. As you begin to log successes, publicize them well.

Make measurement easy by investing in automation. There are two reasons why this is key to a successful measurement program: First, proper automation ensures accurate and consistent measures (this is not to suggest that developers are not honest and accurate, but only that any measurement you ask a person to perform is subject to all of the inconsistencies and errors to which humans are prone). Second, by showing that you are willing to invest in lightening developers' burdens, you are encouraging them to put the proper effort into executing the non-automated parts of the process.

Automation is widely useful. Modifications can be made to a time-keeping system, allowing effort to be tracked at the level required for your measurement program. An automated defect-tracking system can be employed to keep track of testing, defect and rework statistics. Automatic code-counters can maintain size counts on your software. The test suite can be expanded to include tools that extract test time, defect statistics and cyclomatic complexities. Tools can be developed to work with specific integrated development environments to determine complexity or size information about the code as it is being designed and developed.

The trick to making automation work is to identify which of your current tools could be modified to do measurement or collect exportable information that meets one or more of your objectives. A configuration management system, for example, could be modified to collect effort information automatically by time-stamping check in and check out. If you could automate the collection of this information and enforce rules concerning how these activities occur, you could keep excellent track of the amount of time spent doing code and rework on your software systems. Evaluate the types of data your tools collect and how available and exportable that data is. Develop interfaces with other products and stand-alone utilities to build on existing capabilities. Involve the developers in this evaluation — they will have wonderful ideas about areas for which small investments might reap big dividends.

Impersonal Metrics Only

Finally, combat the biggest perceived negative of measurement programs: Never let your measurement program become a basis for individual performance evaluation. While it is important to establish realistic goals for individuals and it is entirely appropriate to use measures to determine the degree to which those goals are achieved, you must take care to minimize any overlap between these evaluations and those that are essential to your measurement program. It is already a stretch to get developers to take the time to provide accurate data; if they believe the information might be used against them, it will become impossible.

There are several effective ways to prevent the impression that measurement programs are being used to evaluate individuals. First, limit the visibility of data that tracks individuals directly. Make sure that all public presentations of productivity and defect data show statistics at higher levels of abstraction. Depending on what makes most sense, do this by software component, systems or even by development team. Remember, you may be measuring output and quality, but you're not recording all of the factors that would explain productivity variations at the individual level.

Software teams enjoy well-planned, well-executed software projects on which they feel they are given sufficient time to build quality systems and that they have a degree of control. They will willingly support programs and processes that take them closer to this goal. As the person responsible for instituting a measurement program, it is an important part of your job to make the measurers understand what they stand to gain from the endeavor. Do this by involving them in the process from the ground up, offering incentives for meeting measurement goals and making sure that you and your management support the program properly. You must also make measurement as easy as possible and convey a clear vision of how it will lead to better plans and processes. When you begin contemplating a measurement program, don't make the mistake of dismissing the measurers as a necessary evil to be dealt with when the program is implemented. Take the time to understand what their needs and concerns might be and address these early on. Involve them throughout the process and treat them as partners in your quest toward better software.

Controversy Continues in the Metrics Arena

Counting lines of code is still popular where function point expertise is lacking.

Though you may think of developers as the stereotypical objectors to metrics programs imposed from above, they are usually less strenuously opposed, say experts, than the managers whose performance will be assessed by their groups' aggregate statistics.

"The biggest barrier to metrics programs is not the individual developers but the senior executives who will be affected by the metrics," observes Capers Jones, chairman of Software Productivity Research Inc. in Burlington, Mass., and author of *Applied Software Measurement* (Second edition, McGraw-Hill, 1996). "Years ago at IBM, I watched as 15 directors of programming, each responsible for thousands of programmers, stood up one by one to protest a new metrics program, claiming that their products were unique and couldn't be measured, until it became clear that they were operating in such a political world that they were afraid that some projects would be found to be better than others. The executive VP saw that this pattern was emerging, and ordered them all to comply. What they found was that, of those projects that ranked near the bottom, some had poor practices and others were indeed quirky, but after a few years, everyone was better across the board."

IBM — once the largest software company in the world — has indeed been a leader in corporate metrics programs, such that today, most Fortune 500 companies are members of the International Function Point Users Group (www.ifpug.org). Smaller companies and start-ups, however, are less likely to institute metrics programs. If you're the first in your organization to look at measuring the quality of your team's output, where do you start?

"If you are measuring only to facilitate quality improvements, you're going to collect different things than if you are trying to track process improvement. You should have a goal first and then pick the set of metrics that supports that goal, keeping in mind the capacities of the measurers and the amount of recurring time you want to invest in measurement," says Arlene Minkiewicz, chief scientist at PRICE Systems L.L.C., a company based in Mt. Laurel, N.J., and offering a suite of software estimating products.

While there are several major types of measures, Minkiewicz advises that "what measure you select is not all that important, so long as you consistently measure the same thing. Measuring lines of code — which is easily automated — is often better than measuring function points — which is better for pure productivity studies — if you have no expert function point counters or don't want to take the time to count function points."

The major metrics, according to Minkiewicz, are:

1. Size, in terms of such metrics as source lines of code (SLOC), function points, object points, the number of shalls in the requirements, the number of weighted methods per class or the number of objects or classes, to name just a few. A fair amount of controversy exists over the validity of the various size measures.

2. Effort hours, tracked at the resource and task or activity level (for example, Joe Smith spent 20 person hours working on the design task for project A).

3. Calendar days, tracked at the task or activity level.

4. Human resources, by resource type at the task or activity level (for example, the number of designers — either peak or average — during coding).

5. Pages of documentation (such as requirements, design documents or specifications).

6. Defects found during various life cycle activities (expressed as defects per unit of size).

7. Latent defects or those released to the public (again, as defects per unit of size).

Of these metrics, function points, developed by Allan J. Albrecht of IBM in the 1970s, have achieved the greatest acceptance. A function point comprises the weighted totals of the inputs, outputs, inquiries, logical files or use data groups and interfaces belonging to an application. Nevertheless, counting logical or physical lines of code — despite the problems inherent in using the technique across different language levels and programming styles — continues to be a popular size metric. And backfiring, a method of extrapolating function points (typically in legacy code) by counting LOC and using a predetermined ratio of function points per LOC for the particular programming language, is really no better than counting LOC alone.

Most of Minkiewicz's current clients, for example, are aerospace and defense contractors who are still primarily dependent on the SLOC metric even as they migrate toward object-oriented systems — where there is even less agreement as to how to evaluate size and complexity.

"The OO community knows less about metrics than the others," says Jones. "The primary disadvantage of object points is that you can't measure procedure points. Using object points for OO systems and function points for procedural systems is like measuring the output of a combustion engine with horsepower but using some other metric for an electrical motor. The advantage of function points is that you can compare different systems and languages. Object points only work for objects."

Minkiewicz, however, disagrees. "Among my clients, I've got pockets of people screaming for something other than source lines of code. These are people with real-time or OO systems who have never used function points, so they need something that can measure those intensive middle-tier processes. Function points are really geared for a database on the back end and a user interface on the front."

— Written by Alexandra Weber Morales

5.5.14 "Habits of Productive Problem Solvers"
by Larry Constantine

Stuck on a problem? Process meta-rules can keep you from wasting time when you don't have time to waste.

One way or another, most of us who work in software development are in the solution business. Our profession is problem solving. Of course, we may think of it differently. For instance, my company specializes in product usability. However, whether we are creating the slides for a training program or designing an industrial automation system's user interface, our real job is solving problems. In our business, we are often constrained by impossible and inviolable deadlines yet mandated to produce breakthrough solutions — on time and within budget. On more than one occasion we have produced patentable innovations on demand.

Whatever you do often and well enough, you eventually learn at a level that goes beyond the boundaries of ordinary technical know-how. The best problem solvers learn the tacit logic of creative problem solving, whether or not they are conscious of the axioms and rules of this logic. Those processes contain patterns that efficiently lead to good solutions, regardless of the nature of the problem or the domain within which solutions are being pursued. After years of building insight into complex problems and producing creative and practicable solutions, conscious practice is transformed into efficient habits — habitual working modes that spell the difference between those who are good at something and those who excel at it.

When asked about the secrets to our successes, I usually reply that there are no secrets; it's problem solving, not rocket science. However, a few years ago I started turning some unconscious habits of productive problem solvers into aphorisms to share with others. These are not so much rules for a particular method as they are meta-rules for effective problem solving in general. We could cast them as "patterns" along the lines established by the so-called patterns movement. They would still be meta-rules, so I won't try to expand them into something else. Because there are so many of these process meta-rules, I'll focus on a powerful subset: those that help keep us from wasting time when there is no time to waste.

Getting Unstuck

Individuals and project teams can get stuck in numerous ways, but most commonly they are either stalled or spinning their wheels. It is important to understand the difference so you can pick the most effective tactics for getting unstuck.

You know you are stalled when you draw a blank, when no one has any idea where to start or how to proceed. Long silences around a conference table, serious doodling that threatens to wear through the paper, or blank stares at a display are all indicators of mental stall-out. Writer's block — which every columnist facing a deadline and every would-be Ellison or Vonnegut knows well — is a prime example, but you don't have to be a writer to be familiar with the feeling. Time slows, the tick of the clock — real or imagined — becomes the loudest sound in the room, and the sense of doom heightens with every passing minute.

When you are stalled in a car, you need to restart the engine. If the battery is dead, you need a jumpstart. The process meta-rule that applies to being stalled is simple.

Start somewhere.   It is easy for software people to become obsessive about starting at the beginning or doing things in the right order. However, when it comes to problem solving, doing anything — even the "wrong" thing — is almost invariably better than doing nothing. Authors who write for a living usually have a whole kit bag of personal tricks for overcoming writer's block. Word processing, owing to the flexibility of the medium and the ease of editing and rewriting, has ushered in an array of new tricks. If you are stuck for an opener, you can write the conclusion first, then construct the arguments that get you there. Or, you can jump into the middle, even plunging into topics without knowing where they will ultimately fit. You can always rearrange them later to form the story you need to tell.

Programming, software engineering, user interface design, and the sundry skills that come into play in producing software all have their variations for overcoming inertia. Despite what the advocates of process maturity and formal methodology will tell you, it is not always necessary that you start at the beginning. It is necessary that you start somewhere. If you fail to get the project moving, it will rapidly become irrelevant that you were stalled in the right place.

In usage-centered design, for instance, the modeling process is supposed to begin with the roles users play in relation to a system. If that draws a blank, however, why not start with something you do know or understand? You can sketch out some tasks that you are confident the system must support. Not only are you doing something, but you will often be inspired to fill in some of the blanks regarding your users.

Because software developers love problems, they often tackle the really interesting and messy stuff first. It's more fun! As a rule, however, progress will be smoother and swifter if you proceed from areas you know and understand and push the boundaries slowly out or back into areas that are less well known or understood.

Begin with what you know. Sometimes you are so lost that you have no familiar ground on which to stand. In these situations, you must create what you know out of thin air by using two preeminent techniques: brainstorming and blankfilling.

Brainstorming, familiar to nearly every citizen of the developed world, is the preferred and obvious way to jumpstart a stalled problem solving process. The freewheeling and uncensored atmosphere of tossing out ideas without debate or discussion is just what is needed. In fact, brainstorming incorporates into its procedures another of the great process meta-rules.

Create before you critique. Separating out the generation of ideas from their review and refinement accelerates problem solving. Criticism, highlighting the constraints and limitations, searching out the problems, and pointing out the downside are all ways to produce good solutions, but when you are already standing still, slowing things down hardly makes sense. Besides, we are all better at critical analysis when we have something to critique. So build the ideas, the proposed solutions, before you start tearing them apart.

I did say there were two jumpstart techniques. Brainstorming is the lively right-brain variant, but blankfilling is a more methodical technique that appeals to our logical faculties. It's like generating your own forms or checklists. To move ahead, you need to start with what you don't know. You can give it structure by labeling two sheets with "What we don't know!" and "What we know!" After the lists are complete, you can either proceed from what you know or start filling in the blanks on the other sheet. Either way, you are no longer stalled.

Circle Game

Watch software development professionals enough and you realize that we have a tendency to spend a lot of time spinning our wheels. Wheel-spinning can be a solo activity or a team sport. You can sit staring at your screen for hours as you go back and forth between alternative ways to resolve some programming problem or you can join with co-workers and turn a half-hour design briefing into a three-hour shouting match. Time spent going in circles does not advance the project. Debates without resolution and discussions that lead nowhere do not bring the team closer to meeting its deadline.

Because so much of my company's work is conducted under heavy time-pressure, we often find ourselves invoking what we regard as the foremost among process meta-rules.

Don't spin your wheels. A car that is stuck on ice or up to its hubcaps in mud gets nowhere fast as long as the wheels are spinning. A project team caught in a go-round over the best way to organize the startup screen is also going nowhere.

You know you are spinning your wheels as soon as you start repeating yourself. When the same arguments are being met with the same counterarguments, when you find yourself reviewing the same materials without new insight or understanding, you know you are spinning your wheels. Under the pressure of modern, accelerated, and time-boxed development cycles, you don't have time to spin your wheels. So don't.

As soon as my team and I recognize that we are spinning our wheels, we stop — and do something else. We move on without resolving the issue if the nature of the problem and the process allows. Or we switch streams and work on another problem or another part of the same problem. Or we switch gears, deliberately speeding up or slowing down the process to get a grip on the problem.

When first learning new modeling techniques or design approaches, groups will often get caught going in circles — without stopping and without progress. When we recognize this state, we draw the team's attention to it and suggest the simple process meta-rule. Suddenly a mental light bulb is lit and progress can begin again.

When you and your teammates find yourselves spinning your wheels, the essential thing is to do something else. What I usually suggest, to continue the automotive metaphors, is to back up, shift gears, or try a different route. Sometimes your wheels are spinning now because of what you didn't do earlier. You need to backtrack and see what you can do to get more information or refine the work. Sometimes the wheels spin because you are going too fast or too slow. In complex problems, you can often achieve progress by ignoring details, skipping over unresolved issues, or practicing that grand technique we consultants call "suitable vagueness." But sooner or later, the details need to be detailed and the unresolved issues resolved — otherwise you keep spinning your wheels in the muck.

Alternatively, you may be spinning your wheels because you are obsessing over minutiae or worrying a problem to a premature death. Then you need to speed up, leap forward over the piddling matters, and move on to the next issue.

Most often, however, the best course is to take another course. Attack a different problem, work on another model, explore an unrelated issue. If you are mired in the mud of trying to construct the narrative bodies of your use cases, you can work on the map that gives an overview instead. If your data model is being churned to death, work on the search strategies and algorithms.

Another time-waster is solving problems that are not part of the problem — especially undesirable when time and budget are tight. "Oh, we never do that," you are probably thinking. What about all those clever features you dream up for the next release, features that are not in the specs and that nobody has requested? The truth is that programmers love to program, so they are more than happy to think up new things to program. Often without awareness, they will invent interesting challenges to be overcome.

Solve problems, don't create them. Some design teams are particularly adept at taking difficult problems and expanding them through creative feature creep into bloated monsters. You may have seen some of their software.

Of course, sometimes the problems are already so enormous and overwhelming in their complexity, that you have no idea where to start, much less what to add. Faced with dead end after dead end, you can become so discouraged that you are ready to give up. That urge to surrender sometimes points to the best way to cope. The last of my process meta-rules could be phrased in many ways — give up, go away, set it aside — but I'll express it in my favored mode.

Sleep on it. I do some of my best work at night — not in the evening marathons that are one of the exhilarating but exhausting hallmarks of running a multi-national operation from a home office, but after I finally go to bed and fall asleep. I have come to trust the workings of all those non-conscious search and retrieval functions, and irrational little engines of reasoning scattered around the brain. Time and time again, I have gone to sleep with an unsolved problem on my mind, only to awake the next morning with a new idea or fresh approach.

Now, I will deliberately turn the toughest nuts be cracked over to that network of little subprocessors while I go about other business or get some sleep. How many times have you struggled to remember a name, given up, then ten minutes later suddenly gotten a mental priority interrupt when the name comes to you? In the interim, your myriad concurrent processing demons have been scurrying around without supervision or awareness to get you the answer.

So, stuck without a solution, I will often assign it to my unconscious. I take a moment to review and restate the problem to myself, then I give it over to background processing. I do not puzzle over it or keep poking at it. I know that sometime in the next day or two I will be about to take a bite of lasagna or will awaken after a good night's sleep and there it will be — the solution or the germ of a solution.

Software specialists are skeptics, for the most part, distrustful of intuition or the unconscious. When working in a project team, I might just suggest that a difficult problem be put on the back burner, that we not worry about it now, but see if anything turns up. It works. In the meantime, we are working on some other part of the problem and making progress there, so we can't lose.

In fact, you probably are facing a tough software problem right now. Maybe you should take a nap.

5.5.15 "From Engineer to Technical Lead"

by Andrew Downs

How to run a focused and efficient team when you're still wet behind the ears.

Congratulations! You've just been promoted to technical lead. Now what? Any promotion can be stressful, but if you were previously an engineer (in title or practice), this represents your first venture into the lowest tiers of project management. In your new role, "soft" skills such as running meetings, communicating effectively, reviewing performance, scheduling and reviewing code will be as important as your technical background — and if you leave the human issues unattended, they can lead to hard problems. What's more, whether you have been promoted from within the organization or not, you'll face the challenge of earning your colleagues' respect.

Starting Off Right

New teams are rarely cohesive or efficient, even if team members have worked together on various parts of the same project. Members have to get to know each other and the team leader to learn who possesses what skills (the so-called hidden organizational chart). Unfortunately, this process takes time, something most projects have little of.

In my experience as a technical lead, I've found that a "kickoff" meeting at the beginning of a critical phase helps my team crystallize quickly and clears up any misconceptions regarding individual roles. If you have ever been subjected to a project that transitions from phase to phase with little or no acknowledgment of the milestones, you'll appreciate the effect of such a meeting.

On a project that I led for the U.S. Navy, I prepared for the kickoff meeting by reviewing the existing documents and writing some additional ones. We had completed a prototype phase, a lengthy analysis and finally a high-level design phase; this resulted in a set of documents describing the system architecture and specific processing requirements for the distributed applications (of which there were three types). These documents stated what needed to occur in the system, but not how it should be done. I knew from experience that developing from high-level documents rarely produced the desired result. Hence, I created a low-level design for a generic, modular distributed application — in accordance with the feedback I had received from seasoned developers — and then e-mailed this document to the team several days before the meeting. This generic model could be easily extended to encompass the specific requirements of each of three types of distributed application. It relied on a plug-in type of architecture, allowing for swapping functionality in and out, though at build time rather than run time.

I told the team that this document, in whatever form it was to become, would be the basis for our work for the next year, and then I allowed everyone to assimilate the proposal. Even if you believe a given design is the best, your team must reach that conclusion on its own. The team's response was astounding: Everyone was familiar with the document. And although it was modified during and after the meeting (thanks to the developers' comments), the core remained the same.

On that occasion, our meeting took a whole day, but kickoff meetings may last just an hour. Make sure to take breaks, and have the company pay for a meal. We also invited teams who working on other aspects of our project; this helped lay the groundwork for interfaces between the various software modules.

As for the meeting's structure, I began by stating the project goals in the form of requirements — you could also review the high-level design or architecture — before I moved into the low-level design. During the meeting, the team critiqued and refined my ideas, which allowed them ownership of the process. Your team's critique should take about half of the allotted meeting time. After that period is over, review and record the agreed-upon design. Then assign individual tasks, keeping in mind that tasking should take no more than one-fourth of the total meeting time. (A side benefit of tasking in this semi-public setting is that each team member will immediately know who is doing what.)

To wrap up, I reviewed our accomplishments at the meeting and had each team member restate his specific tasks. This "briefback" reduces confusion and limits the number of follow-up questions. Obviously, your team needs to be comfortable asking questions at any time, but it's more efficient if they ask their questions during the meeting rather than after the project is underway.

Kickoff meetings are always a good idea, even for a team of two assigned to a one-week project. Not only will you keep your team leading skills sharp, you'll also have an audit trail (that is, if you remember to take notes). You should never have to ask "What did I do that week?" And since small projects have a habit of ballooning into large ones, taking some extra care early on may reduce your stress level if the project continues longer than anticipated.

Keep Everyone Informed

In the late '80s, I enjoyed reading the daily space shuttle status reports on a NASA mailing list. Each report recorded that day's activities and outlined what was planned for the next day. I had fun interjecting into random water cooler conversation comments like "Hey, did you know they just moved Columbia to the vehicle assembly building for mating with an external tank?" (Yes, I received some strange looks.)

I wanted to create a similar report for my team, but make it weekly instead of daily. I adopted the same matter-of-fact tone I had seen in the NASA communiques:

From: Andrew
To: Communications team, other first-line leaders
Subject: Communications team status report for 12/12/99
Tasks completed:

- Joe wrote the low-level thread interface API.

- Ivan on vacation.

Work planned:

- Joe will re-implement the application event loop using the new thread interface.

- Ivan will assist with unit testing of database update module.

My (status) report included the tasks completed in a given week and the projected tasks for the upcoming week. It also assigned tasks to specific developers. I didn't specify how long a task took to complete because we tracked our hours through timesheets; however, adding man-hours would be a good way to measure productivity.

Your boss should also receive your status report, so she can follow your progress. A good boss will query you about tasks that keep appearing on the to-do list for the next week, but are never moved to the "completed" section. Be ready to explain why a particular task is perpetually unfinished, or move the task to an "unscheduled but required" category. Just don't forget to address it later.

More Meetings?

Depending on project length, status meetings can be informal and infrequent. For our large project, weekly one to two hour status updates involving all team members were appropriate. The weekly e-mail reports formed the basis for discussion.

The final encounter should be what is affectionately termed a "postmortem." Whether you hold this analysis meeting (also known as a retrospective) to discuss the entire project or just a particular stage, make it a positive experience. Ask questions like: What worked? What did we learn? What would we do differently? Everyone will probably have different responses. Write them down.

An additional form of feedback may not immediately impact the project, but will provide future help: Insert yourself into the review process for each team member. At a minimum, summarize each member's accomplishments and send it, well in advance, to whoever conducts performance reviews. (For an annual review, submit your input a month in advance.) Follow-up to make sure the reviewer received it, then lay off: You've done your job.

But I Want to Code!

You may still write some of the code, but your role as technical lead is to ensure that the rest of the team writes code that efficiently fulfills the project requirements and matches the appropriate level of design.

Scheduling and conducting code reviews is an essential part of achieving that goal. Code reviews keep people honest and concerned with the quality of their code. Plus, a second or third set of eyes on a piece of code will always uncover things the author can't see. Follow up on problem areas at subsequent meetings, so that your team knows how important the reviews are. For small teams, you can conduct code reviews with just a few formal roles:

- The moderator should ensure that everyone attending the review receives copies of the code well in advance. He or she also assigns a reviewer for each code item (file, routine, and so on.)

- The reviewer should be a senior developer if possible, but this can be used as a training exercise for experienced junior developers. (This is one way in which you train them to become senior developers.) The reviewer should have a copy of any design documents that pertain to the code item under review.

- The author may walk the reviewer through the code and clear up any obscure items.

- The note-taker records the disposition of each item. On our project, we filled-out a simple status form for each item, stating when and by whom it was reviewed, any problems found, and if it passed or required revision (and subsequent review).

We often only pay lip service to documentation, even though most would agree that any project will suffer from poor documentation. Good programmers who write well and enjoy writing something other than code are rare. If you have some of these creatures on the team, hold on to them! If not, write the documentation yourself.

Grow the Team

Good team leaders give everyone a chance to excel. I tend to task others conservatively, which means some people have to ask for additional work. However, you shouldn't be afraid to give your team members more work to see how they react. Also, assigning tasks becomes easier once you understand another's style. Provide tutorial — in the form of documents or short courses — for both rookie and veteran team members. Consider including nonteam members, because sharing information (especially high- to mid-level design details) can result in greater overall design efficiency.

On my project, at least one half-day overview session (with numerous mid- and low-level design details) was attended by other teams, most notably the user interface team. This eventually led to a simplified system design: Once they were able to peer inside the "black box" that comprised our background applications, the interface team members were able to streamline both their design and ours.

You should also assign more complex tasks to senior developers and, perhaps, have a junior developer assist. If practical, assign the task to the junior developer, and have a senior developer act as a mentor. The mentor can help refine the design or provide a suitable code skeleton with which to work. Let the senior review and refine, but not write, the additional code.

Find someone who is both senior and an excellent developer to become your second-in-command when you're not available. He should also be a sounding board for any new ideas you may have. You may not always see eye-to-eye, but you should at least agree on what tasks need to be done in the short-term. Everyone has a unique personality and style; don't expect a clone. Focus instead on the results attained by your partnership.

Play Well With Others

Since many projects require multiple teams, you will need to occasionally run interference. On a large project, you may be the official liaison between your team and other players that include outside companies or contractors. To effectively defend your team's interests, you must know your team's responsibilities. I found that a statement of work (SOW) is useful for this purpose, but it must be written in very specific terms. If a SOW is not available, the system architecture and design may dictate or constrain your team's responsibilities.

Remember, you may have to say no to another team's requests. Don't refuse in an arrogant or malicious manner, or others will repay you in kind. Instead, explain why something can't be done. Relate it to time or money constraints if possible; this always gets attention.

Take This, Brother...

Handing off part or all of a project to another team can be difficult because many developers become attached (intellectually and emotionally) to their code. How can they trust their creation to someone else? The transfer of responsibility can also be a political issue: Another vendor may offer the customer additional modifications (using your source code as a starting point) at a lower rate than your company charges, or the product may be modified by the software maintenance team. Either way, the project and code belong to the customer, unless otherwise stated in the contract. Do your best to assist the new group, without compromising your team's objectives or performance.

You can successfully move from engineer to technical lead. And although you're leaving behind the carefree days of writing code, your new role will allow you to have a greater impact. As technical lead, you will be able to directly influence the work of others; junior and senior developers will rely on you for design and implementation guidance. And whether the project is large or small, applying "soft" skills will ensure that your team is focused and efficient, while preserving a record of its actions for later analysis.

Favorite Books

Here are a few books I have found both useful and enjoyable as a technical lead:

- *Object Solutions* by Grady Booch (Addison-Wesley Publishing Company, 1996). Read this book first if you are involved in an object-oriented project. If not, try to read it anyway. It describes the development process at multiple levels and contains many useful tips.
- *Code Complete* by Steve McConnell (Microsoft Press, 1993). Most developers probably buy this book for its treatment of code techniques, but it also discusses team development, code reviews, testing and personality. The author provides many additional references.
- *Show-Stopper!* by G. Pascal Zachary (The Free Press, 1994). This is the saga of the creation of Windows NT. It reads almost like a team lead's diary. The story is very intense and personal, reflecting the same types of conflicts I experienced (and which you may also experience).

— Andrew Downs

5.5.16 "Effective Resource Management"

by Susan Glassett

People stay with jobs they enjoy, even when they can get paid more elsewhere. Keeping your staff educated and interested will maximize productivity — and employee retention.

While I finish this article, I'm sipping a beer and watching my development and sales teams play pool, eat, and drink in celebration of our latest product release. I'm struck both by my team's accomplishment (a complete suite of enterprise products in less than a year) and by our luck. If we hadn't recruited just one key team member or if we'd lost a senior engineer, this product never would have gotten out the door. In addition, if anyone had failed to perform his or her assigned tasks, this release would have been delayed. There was no room for error in recruitment, retention, or assignment of my software development team.

This is the state of software development. As companies continue struggling to leverage information technology to create strategic advantages, the demand and importance of software projecting deployment has increased. Yet, depending on who you wish to quote, 40% to 60% of all software projects will be canceled or result in systems that fail to deliver their intended value. Despite tool, technique, and environment advances, software development remains a labor-intensive activity and the resource pool is not keeping up with demand. In this type of hiring climate, recruiting and retaining personnel is nearly impossible.

Tools, methodology, and management improvements increase, but nothing affects productivity more than getting the right people to do the right work at the right time. This is what resource management is all about.

Effective Personnel Management

If you want to keep up with the current marketplace, you must recruit highly skilled and experienced staff. The challenge is retaining them and matching their skills, experience, and desires to the work that must be performed. Assigning work is crucial to recruiting, retaining, and maximizing team productivity. This means matching developers' skills and experience to project needs, finding work assignments that interest them, creating compatible project teams, and successfully predicting both short- and long-term personnel requirements.

Matching your staff's skill and experience to project needs is multidimensional. The better the match, the better the productivity. Whether they are technical or applicable business skills, such as accounting, customer service, order processing, or inventory control, skills are not binary. There's a gradation of proficiency, which you can represent with a linear scale mapped to your project's minimum requirement. Proper experience-matching will affect individual productivity, but it has more to do with risk. Experience is insurance. The more risky a particular project aspect is, the higher the requirement for someone who has done it before. High-priority, high-risk projects need the most experienced teams.

A software development professional's skill, experience, and desire to accomplish the task determine his or her productivity. You can quantify skill and experience, and your staff can't do the work without them, but these two alone aren't enough to achieve peak performance. The team member's desire to work on the assignment is just as important. A C++ programmer who has taught himself Java will be more productive than a Java programmer who doesn't like the language. The take-home message: pay attention to what people want to do. It can make all the difference.

There is a difference between individual and team productivity. Software products need to be built, tested, and maintained. To achieve that goal, you need a functioning, coordinating team. I have managed organizations with productive individual software developers who couldn't work as a team. Along with the right skills, experience, and desire, your team also needs the right chemistry.

Leadership and mentoring are crucial to running a well-oiled software development machine. Many organizations find that smaller teams, working on individual components of software solutions, are more effective than larger ones. This increases the importance of having strong technical leads or team managers.

Effective resource management includes skills and experience capacity planning across an organization. You can't recruit "just-in-time" for projects in today's job market. To have the right skills and experience available, enterprise-level project planning is necessary. You should assemble and track a portfolio of project needs, schedules, dependencies, and relative priority. This portfolio must include current and planned projects and staff assignments, and will also focus your recruiting efforts.

Resource capacity planning also provides for contingency planning. Even when you do everything right, the unexpected can happen. Births, resignations, and marriages can wreck havoc on the best-laid project plans. I recommend identifying critical staff and asking yourself, "If Sally won the lottery and quit her job, what would we do?" It is important to line up and groom backup people for every critical role. These personnel often come from the same project team, which lengthens the project's schedule. You might also find backup staff in lower priority project teams. In either case, your contingency plan should include canceling or postponing a lower priority or dependent project.

Effective resource management must include recruitment and retention. Retaining people who are challenged and happy at work is the most important element of successful recruitment. People choose to work where there are other people with whom they like to work. People relocate where colleagues assure them they can work on challenging, career-advancing, and enjoyable projects. People are attracted to companies where other people seem to be enjoying their work, even when they can get paid more elsewhere. Aligning someone's skills, experience, and desires with the activities performed is very important to employee retention.

Recipe for Effective Resource Management

Effective resource management is ongoing. The following activities are iterative and often performed simultaneously. I've presented them in the sequence that works best for me.

Set up a resource inventory. To make the best assignments, you need an inventory of your staff's skills, experience, and desires. Your skills inventory should include technical skills and business functions that relate to your organization's projects. Experience records should also be as detailed as possible. For example, recording "two years programming in Java and five years programming in COBOL" is preferable to "seven years programming." Interest or desire is harder. I use simple comments to record interest in certain projects, business areas, or wanting to work with certain people. I use the lower end of the proficiency scale to record interest in acquiring a skill.

Record your staff's proficiency. For each skill, proficiency should be recorded. A common proficiency scale is a simple rating of 1 to 10. Ideally, managers know the skill levels of the people who report to them. In technical disciplines, this is not always possible. Some organizations create standardized tests or use outside consulting firms to perform the tricky tasks of assessing certain proficiencies. I have found that if I ask several senior technical people who have worked with the employee to rank his or her skill in a given area, there is normally a convergence that lends itself to simple averaging. When there is no convergence, I must investigate further. Precision isn't as important as getting a rough range of capability.

Build project plans. Project plans require good project management, in the sense of planning and estimating so you can declare, understand, and schedule individual tasks. Your project plan defines the staff assignment requirements. It should include a schedule of activities, skills, and experience for different roles, and the relative priority of the project and its deliverables.

Make assignments. Match up your resource inventory and staff availability with your project plans. This is not an algorithmic activity — there's no replacement for management intuition. I use the availability, skills, and experience requirements to come up with a list of candidates, then use my knowledge and notes to make a choice. This is an ongoing process. Building teams and determining assignments is easy when you have a few projects and several dozen people. As the numbers increase, this activity's complexity makes it next to impossible to perform well without automated tools. I personally reached that point when I was tackling about 6 to 10 complex projects with a resource pool of 50 to 100 people.

Review the project teams and assignments. After I make a set of assignments, I like to step back and look at the teams. I find that using several smaller teams provides maximum flexibility. I may break a single product or project into a half dozen separate component-oriented teams. I like project teams of two to three people with a strong senior technical person as a designated team lead. This isn't always possible, but it's a reasonable goal. Chemistry and scheduling adjustments are easier when projects have several, small deliverable-oriented teams. Ideally, you can arrange work assignments so that each small team delivers something tangible on at least a monthly basis. This allows for accurate progress tracking. At this point, I also like to perform a reality check on the assignments. I insist on weekly time tracking and calendar updates from all staff. This lets project managers match assignments to individual calendars. If you assign a task requiring 75 hours of work in two weeks to an employee with 3 vacation days planned, it's not going to get done on time. The more real the schedules, the more predictable the outcome.

Perform capacity and contingency planning. You should build project plans and make staff assignments for any project being considered for the future. This lets you predict staffing needs so you can plan recruiting and staff development. I also like to do "what if" analysis against my backup plans for losing a key employee. This helps me determine my risk areas so that I can attempt to mitigate them.

Admittedly, the above recipe involves a lot of detailed planning, tracking, and analysis. It takes time and commitment. Nevertheless, any software development organization that intends to satisfy business demands must manage its staff assignments at this level of granularity.

5.5.17 "What's Wrong with Software Development"

by Christine Comaford

---

*Examine nine problems of software development and learn
how to avoid the pitfalls.*

Software development is in a worse state today than it was five years ago. Thirty-one percent of U.S. software projects are canceled before completion, costing $81 billion annually; 53% of projects overrun their cost estimates by 189%, costing $59 billion annually; and only 16% of projects are considered successful! The Standish Group of Dennis, Mass., recently revealed these shocking statistics.

What's the answer? We need to know why we're building an application and how to build it. We need a roadmap of proven processes. We need best practices to prevent us from repeating the same mistakes; reduce project cycle time and costs; increase software quality; and retain, maintain, and distribute expertise without depending on transitory experts.

According to the Gartner Group, 75% of the five-year ownership cost of client/server systems is in labor — and labor costs are going up, while hardware and software costs are going down. Many companies spend up to 10% of a developer's salary on education — yet when the developer leaves, the expertise goes with him or her. People are getting smart, companies aren't! Let's look at a few problems in software development and consider how best practices will solve them.

1. We need to better understand our business.

We don't understand how to interact with executives, prioritize or cost-justify projects, and secure sponsorship. Solution:

* Attend user groups: the Society for Information Managers and Corporate Association for Microcomputer Professionals are great, both headquartered in Chicago. The Deciding Factor, a group in Basking Ridge, N.J., will give you a project prioritization spreadsheet for free.
* Explain the purpose of your business in one sentence. Then make sure your Information Systems priorities match it. This is also helpful in ranking application features if you need to cut some.
* Decentralize development. Developers must be in business units — this is the best way for them to understand user's needs!
* Be brave. Do post-project reviews, using a standard form.

2. We need to better understand our users.

They are powerful and are only becoming more so. They have new responsibilities: to help us gather requirements, assist in prototyping, and do incremental acceptance. On average, only 61% of originally specified features and functions are available in the final application (Software Productivity Group Inc., 1995). No wonder they don't like us. Solution:

* Do user and task analysis, using a standard form.
* Gather requirements (do four-hour sessions per day), using a standard form.

- Do at least one session before you start prototyping.
- Use a standard prototype feedback form to ensure consistent feedback is gathered on each revision.

3. We need to better manage our development staff.

Teams are too big; we don't know who has what skills; skill sets and project needs are rarely matched. Solution:

- Keep a skills inventory (with roles, tasks, skills, training).
- Use small teams of six to seven people.
- Have integrator roles (to ensure consistency across GUI, code, business rules, and interfaces to other systems).
- Offer incentives! From 5% to 30% of a developer's potential annual income should come from incentives. Base them on reusing others' code, creating code that can be reused, assuring code quality, and meeting the project deadline.

4. We need to create and maintain an application development infrastructure.

In most companies, each project is an island that does not benefit from any past projects. Maintenance takes up 63% of all software development work. The infrastructure is the best way to reduce maintenance because it lets you document systems better, as well as utilize more reusable components. Solution:

- Set standards. As a minimum, have GUI design, naming, and coding conventions.
- Have predefined best practices for defining, designing, coding, and testing an application.
- Have an application development support team (3% to 5% of total development staff) chartered with making all developers more efficient and informed.
- Assign an object librarian and set up a repository of GUI templates, code, business rules, and data objects.

5. We need to understand what RAD really means.

It doesn't mean fast screen painting. It's concurrent engineering, prototype-driven, reuse-intensive development. Solution:

- Apply RAD techniques to both small and large projects.
- Divide a large project into smaller, concurrent subprojects.
- Use consistent processes.
- Concurrent engineering and controlled prototyping builds fast applications; user-centric design yields business-oriented apps, layered architecture makes versatile apps, proper partitioning of data and process ensures robust apps.

6. We need reliable estimation techniques for client/server.

We lack guidelines and have a poor feed-back loop, if we have one — so we don't compare budgeted results to actual results and thus don't learn from history. Solution:

- Have developers track their own actual task duration, tying one or more tasks to a deliverable.

- We spend too much time on certain activities. Here's a sample timeline I recommend: two weeks for requirements clarification and preliminary prototype, which yields the initial project estimation; six to eight weeks for the real prototype, which yields the initial estimation adjusted 15%; 10 weeks for fleshing out the prototype, adding error and exception handling, tuning code and data access, which yields the completed application, functionally tested; four to six weeks for user profile, integration, and system test. Ready to deploy!
- Determine the number of components (windows, tables, functional areas, queries) and their degree of complexity. Multiply these by the component plus complexity-level weights.
- Adjust estimation by degree of standards, reusability, transaction load, usage, roll-out, performance requirements, administrative overhead. Then staff the project based on complexity.

7. We need to control prototypes better.

We need to time box, schedule, scope, and manage the prototype process. We also need to manage consumer involvement to ensure getting timely (and appropriate) feedback. Solution:

- Do evolutionary prototypes.
- Schedule iterations and time box them. Decide in advance what features will be in what iteration. Schedule appropriate users to be at a prototype and feedback session. Have users, developers, and managers sign off on the schedule.
- Close the "user feedback door" 24 hours after releasing prototype iteration.
- Use a standard prototype review form to ensure consistent user inspection of prototypes.

8. We need to build better architectures.

I've seen far too many flimsy, tactical systems! Without a solid architecture beneath that sexy GUI, we cannot scale our application to accommodate more users, varying data sources, and greater data volumes. Solution:

- Do performance modeling before you write any code.
- Design for scalability; aim for the worst case — your workgroup application being used across business units!
- Use a layered architecture for optimal tuning, performance, and reuse.
- Use predefined, pretested architectures.

9. We need to start testing.

Every application must be subjected to consistent quality assurance, usability, and performance testing methods. Solution:

- Perform testing by a central team — you'll get consistent, high-quality testing.
- Use the helpdesk staff as an additional testing resource.
- Testing occurs much earlier in the development process than it used to — as soon as a component is complete. Development must do functional testing prior to checking code in. The quality assurance team does unit, integration, user profile, and system test. Remember: user acceptance testing must be done incrementally!

- Create a methodology for designing test cases, suites, and scenarios to ensure consistent quality across all your applications.

The "silver bullet" tool doesn't exist. Forget it. Best practices — and not random, non-repeatable processes — will finally get software development on track! n

5.5.18 "Scaling Up Management"

by Larry Constantine

The challenges of large-scale development projects can offer valuable lessons. Looking at the big picture can help you manage a software project of almost any size.

When I was young, I was often told to think big. Nowadays, we are told to think small.

As recent columns in this forum have attested, we are working in the era of the small, select, high-performance project team and rapid iterative development on compressed time scales. Managers are counseled to assemble elite teams of a few highly skilled developers and to scale applications down to be completed within 90- or 120-day time boxes. Projects bigger than six developers working for six months are doomed, according to current thinking among the gurus of smallness. Best to avoid such projects altogether.

You, however, may not have the same freedom to pick and choose that some industry leaders enjoy. You can't neatly dissect every project into digestible six-month chunks. Some enterprise information systems are big no matter how you look at them. You may have inherited a team of 200 and a mandate from top management to use them well. Maybe you just finally got that big break on a really big contract.

Even if you are only part of a three-person team cranking out intranet applications, you might learn something from thinking big. Just as the insanely short delivery cycles of web-time development (see Dave Thomas's "Web-Time Development," in this chapter) can teach us about making effective use of time even when we have more relaxed schedules, the challenges of large-scale projects offer valuable lessons even when applied "in the small." In short, there are many reasons to look at the big picture of large-scale development.

You will note that I have nicely sidestepped the issue of what constitutes large-scale software projects. How many programmers make for a big project? It's all relative, of course. To a manager experienced in leading a technology point team of three or four programmers, coordinating a department of nine might seem overwhelming. In contrast, to someone schooled in the massive scale of such historic and fabled efforts as IBM's OS/2, the 200-developer team that created Windows NT might look downright lean and mean.

Some managers say a large project is any project that is sufficiently bigger than anything you have done before, so that you have no idea how you are going to manage it. This view goes to the heart of the matter. The issue for the software manager is not the absolute size of the team or the source code file, but the effect of scale on the problems of management.

In this column, I want to explore some of the ways that key issues of technical teamwork and technical leadership change with scale, and what this might mean for software projects of almost any size. The question of scale is partly a matter of numbers, such as the number of people participating in a project, but a large part of the question is really about complexity. Whether or not your development team is growing, odds are your applications have become increasingly complex. In either terms, the issues stand out in sharper relief when the scale goes from modest to enormous, so there can be value in looking at a humongous project.

Millennial Management

If you want to learn about technical teamwork on a truly grand scale, I urge you to get a copy of the five-part public television series "21st Century Jet." This joint British-American PBS production is a must for anyone seriously interested in the art and craft of technical management. The video is the story of the making of Boeing's 777, hailed as a triumph of both engineering and engineering management. It is not only a dramatic and engaging tale, but there are many lessons for software development managers. PBS sells the five videos packaged with a book of the same title in the "Business and Finance" section of its web site (http://shop.pbs.org). It could be the best $110 you or your company has ever spent.

No one would doubt that the "triple-seven" team, led by Boeing's Phil Condit and Alan Mulally, was big. Consider a team of 10,000 people scattered around the globe. Mulally and his management team ensured that these people were more than just a team in name. Boeing management was convinced that to achieve a real breakthrough on a tight schedule and budget, they would need teamwork of the highest order, which required radical restructuring of the way their team worked together. Their approach had many aspects, but the pieces that seem particularly relevant to complex software projects are:

- Face-to-face team building
- Consensus-based collaboration
- Interface management through models
- Design reuse.

Building Teams

The highest performance is achieved when a mere group is transformed into a real team, with a strong rapport and shared sense of purpose that can only come from working closely together. Tom DeMarco and Tim Lister call such teams "gel-teams," because they have gelled into a cohesive working unit. Whatever you call them, teams that draw on a shared history and a common culture of work are simply more efficient and effective. If the team's reservoir of shared values and experience is not already full, it must be filled through some deliberate process, and that process is team building. Whether it takes place at an organized retreat or is shoehorned into brief remarks and introductions at a kickoff meeting, team building is compressed experience that helps build a shared history and culture.

Team building becomes more crucial and more challenging as the team size increases. A sense of teamwork that seems to flow naturally from working closely together with a handful of colleagues may have to be deliberately cultivated when the handful grows to fill an auditorium or beyond. Yet, how do you cultivate a sense of "teamness" when there are thousands of people on your team?

I have been teaching team-building and high-performance teamwork techniques for many years, and for all those years I have also been saying that to build a team there is no substitute for face-to-face contact. No matter how large your team or how widely it is dispersed, it is imperative that everyone gather together in one room at least once during the life of the project. This statement always triggers a chorus of yes-buts: Yes, but there is no time for team building on such a tight schedule. Yes, but it's impossible with 150 programmers and designers working in three different states. Yes, but the travel expenses would break the budget. The litany of excuses is endless.

The people at Boeing took the maxim of team building literally and allowed themselves no excuses. Boeing organized the triple-seven personnel into individual design-build teams, each of which was jointly responsible for the architecture, engineering, and construction of an entire subsystem. However, they also gathered all the members of the entire triple-seven team from around the world, assembling nearly 10,000 people in one place. They did this more than once in the project. There is only one space in the entire Boeing empire large enough to hold that many people, an assembly hanger said to be the largest building in the world. Some of the scenes from these meetings are awe-inspiring. This was no crowd of soccer fans cheering on their team, this was the team!

Perhaps it is part of our deeply rooted mammalian heritage, but there is no substitute for meeting face-to-face. Not without reason, important deals are ultimately sealed with a handshake, even when hordes of lawyers have already created reams of signed documents.

The truth is that even for so-called virtual teams of developers sharing work over the Internet and meeting by video conferencing or for geographically dispersed multinational teams spread across 11 time zones, face-to-face contact as part of a team-building process is invaluable. Or perhaps I should say, it is especially invaluable for such teams.

Consensus

There is a telling moment in the original broadcasts, just after the 777 management team finishes wrestling with a tough trade-off. They must decide between using an exotic lightweight alloy that develops harmless but visible cracks or taking a significant hit in take-off weight using conventional alloys. Mulally emerges from the meeting and remarks that this consensus decision-making really works. In fact, the building of consensus and the full involvement of people at all levels in the technical decision making was a cornerstone of the 777 project management approach. Each design-build team had to meet its own schedule and objectives, putting a premium on incorporating construction input into engineering and carrying out tasks collaboratively and concurrently.

Although it's more common to see large teams marked by top-down dictation of design and business decisions, consensus-based decisions have real advantages as project scale increases. Not only does consensus make fuller use of skills and expertise leading to better decisions, but it functions as a form of continual team building that assures higher levels of commitment and participation from all team members. As an individual team member, I see the work as mine because I agree with the direction taken and can see how my contributions have been incorporated.

Interfaces and Interaction

One of the technological innovations of the 777 project was a comprehensive integrated computer-aided design (CAD) system that modeled virtually every feature and aspect of the aircraft as it evolved. It's easy to see this computer system as merely part of the engineering support infrastructure, but it also served important management functions.

As the number of people on a team or the number of pieces in a product increases, the number of interfaces and potential interactions rises even faster. Among 10 people, there are 45 possible interfaces; among 10,000 there are nearly 5 million. In a tight-knit group, project communication and coordination can be accomplished just by raising your voice enough to be heard in the next cubicle. It is another story altogether when thousands of engineers and designers must keep each other posted on work progress and the consequences of particular decisions. What seems obvious and natural in the small scale must become deliberate and structured in the large.

In addition to direct lines of communication and interaction among different 777 sub-teams, the computer models became a common reference point for managing the interfaces among the myriad activities in the design and configuration of the thousands of mechanical, electrical, hydraulic, and other subsystems. Every decision, every solution, every modification was reflected in the common models, which made the impact on other parts of the project and the product apparent. Widely separated groups were connected through their interactions with the computerized models.

Another telling moment in the 777 project arose when a sensor could not be installed within a wing because a fastener was already in that spot. The crisis actually highlighted the importance of the CAD models as a tool for managing the interfaces. The problem occurred because the fastener was not in the computer models, which did not represent details down to the level of individual fasteners. Expensive, late redesign might have been avoided through more thorough modeling.

This process of coordination through interaction with development artifacts can likewise be seen in some large software development efforts. More than 10,000 programmers world-wide have been involved in the creation and maintenance of the one-and-a-half million lines of code in the current version of Linux. It is the source code itself that constitutes the meeting ground for these contributors. Of course, an even better comparison would be a large software project coordinated through a single set of integrated design models rather than connected by the final code.

Design Reuse

The 777 engineers avoided making more work for themselves and for the actual builders of the aircraft through reuse. For example, although each of the eight doors on the aircraft was somewhat different, they were designed to use a maximum of common subassemblies and components. Reuse of designs not only results in fewer unique subassemblies or components to design and build, but it also means there are fewer interactions to be accounted for and tested.

Shared design models can promote reuse by making design problems and current solutions visible across a project. I suspect this is even more true in software than in hardware, because of the greater homogeneity of the basic software components.

Flying Software

You may be tempted to dismiss the 777 experience as irrelevant to software engineering and management because the 777 is an aircraft, not a program, but the real picture is more subtle than that. The 777 is a so-called fly-by-wire system, meaning that the actual control is accomplished by computers that mediate between the cockpit and the plane itself. These complex navigational, control, and display systems had to be programmed, so the 777 story is partly a software engineering story. In fact, a colleague recently told me that insiders describe the 777 as two-and-a-half million lines of Ada flying in close formation — not so much an airplane as a complex software system packaged and shipped in an airframe.

Boeing managed the development of those two-and-a-half million lines of Ada with the same engineering discipline and management finesse as the rest of the project, but there is an ironic footnote to the story. At the insistence of the major customer, the entertainment sub-system was subcontracted to an outside group of software developers. Schedule slippage and the unreliability of that software almost delayed the delivery of the entire plane.

My own flying experiences attest to the difference between the kind of disciplined, team-based software engineering that hardware companies like Boeing practice and the kind of semi-organized hacking that passes for programming in many other parts of the industry. On five of the last eight trans-Atlantic flights I've made, the software-controlled in-seat entertainment system has failed catastrophically at some point. Fortunately, the fly-by-wire software functioned without a hitch and the plane always landed just fine.

Chapter 6

Best Practices for the Environment Workflow

Introduction

The purpose of the Environment workflow is to configure the processes, tools, standards, and guidelines to be used by your project team. As you see in Figure 6.1, which depicts the enhanced lifecycle of the Unified Process, most Environment workflow effort occurs during the Inception and Elaboration phases. The Environment workflow encompasses several major efforts, including the tailoring of your organization's software process to meet the unique needs of your project, choosing the standards and guidelines that your project team will follow, and selecting the tools that your team will use.

What are some of the best practices that your project team can adopt with regards to the Environment workflow? First, recognize that you do not need to reinvent the environment wheel. Look for existing tools, standards, and guidelines within your organization, or better yet, within the industry. Your job is to develop software, not to develop infrastructure. Second, let common standards drive your tool selection. A common mistake is to choose tools first and then get stuck with their proprietary approach to development. If you want your destiny to be defined by your tool vendor, that's your business; we'd rather determine what we want to do and then find tools that best match our vision. Therefore, you should choose your modeling standards *before* selecting a CASE tool and set your user interface design guidelines *before* picking a development tool that supports them. Third, define your environment early (which is why it's a crucial part of the Inception phase). When the project environment is not defined in time for construction, you risk a false start to your project — the use of the improper tools, guidelines, standards, and so on can require the development team to do rework later in the project.

Figure 6.1 The enhanced lifecycle for the Unified Process.

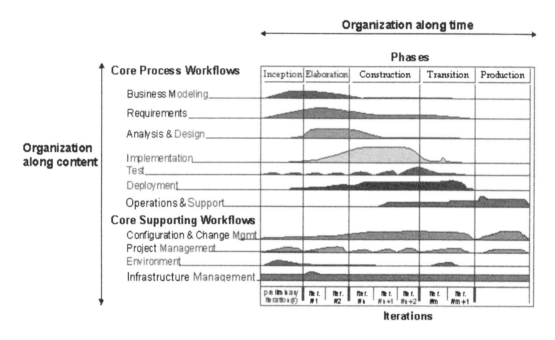

Do not reinvent the environment wheel. Let standards drive your selection, and define your environment as early as makes sense.

6.1 Selecting and Deploying the Right Tools

Software process, architecture, organizational culture, and tools go hand-in-hand. If they are not already imposed upon your team by senior management, you will need to select a collection of development tools such as a configuration management system, a modeling tool, and an integrated development environment. Your tools should work together and should reflect the processes that your team will follow. In *"The Ten Commandments of Tool Selection"* (November 1995, Section 6.3.1), Larry O'Brien presents best practices for determining which tools are right for your project team. O'Brien's philosophy regarding tools is straightforward — "tools, without the right processes, are worthless, just as processes without the right tools are worthless." O'Brien points out that once you have chosen a tool, you should commit to it and continue to use it, even when its competitor brings out a new release because it is unlikely that the new release provides benefit enough to warrant the expense of changing tools. He also stresses the importance of choosing your tools based on requirements — if you don't know what you're shopping for, then why are you shopping? Another of his many gems is that you need to understand how your selected tool will affect your delivered product. For example, some integrated development environments (IDEs) make it easy for you to write code that is locked into a specific operating system — a side effect you may not want.

Select your tools wisely.

Karl Wiegers, in "Lessons Learned from Tool Adoption" (October 1999, Section 6.3.2), argues that selecting and installing a software tool is easy, but making effective tool usage integral to your group's software engineering practices is much harder. His experience is that a tool is not a process, but instead, a tool merely supports a process. He believes that:

- You should be able to articulate the overall expected benefits of a tool to the project team, the organization, and the customers.
- The benefits from most tools come from improving the quality of the delivered product.
- You must weigh the price of the tool against the costs of not using it.
- A cultural transformation takes place as a team evolves from manual methods to the discipline of using defined processes and tools.
- The best chance for successful tool adoption comes if the tool can attack some source of pain your team is experiencing.

Tools are not processes. Tools support processes.

In "Timing is Everything" (October 1996, Section 6.3.3), Roland Racko presents a collection of lessons learned regarding tool selection and deployment. Like O'Brien, he points out that today's "best-of-breed" product is quickly eclipsed within months by a competitor's latest release, and that instead of choosing the best-of-breed based on a magazine article, you should instead evaluate potential tools based on your own specific criteria. Use the magazine articles to narrow your scope, but still evaluate the tools within your own environment to make the final selection. He also points out that your tools must reflect your methodology/process. More importantly, he stresses the ramp-up time needed to learn a new tool, you can't put a box on someone's desk and expect them to be proficient with the tool that very day. This article is worth reading for anyone who is defining the tool environment for his or her project team.

Don't forget the time it takes to learn your new tools.

6.2 Deploying Your Software Process, Standards, and Guidelines

In the Unified Process, the definition of the tailoring of your software process is called a *development case* — the creation of which is well-documented in *The Rational Unified Process* (Kruchten, 2000). But how do you share your development case within your project team? In "Improve Your Process with Online 'Good Practices'" (December 1998, Section 6.3.4), Karl Wiegers, author of *Creating a Software Engineering Culture* (1996), tackles this question head on. He describes the experience of one company that addressed its process documentation challenges by creating a corporate-wide, intranet-based collection of good software engineering practices. In the past, these processes existed only in hard copy form, which made it difficult to share the contents across the company or search for specific items. This company had an interesting notion: it was interested in "good" practices, not "best" practices. They believed it begs the question of who determines what is best and what the reference points are. Furthermore, they (and justly so) expected the initial collection to be improved upon over time, as project teams gained experience developing and applying new processes and created better examples. Therefore, they focused on building a collection of useful, but not necessarily perfect, artifacts that would be better than the resources presently available. A second notion they held was that less is more — leading them to trim example project artifacts down to a manageable size that simply illustrated the structure, contents, and style of the artifact. Their belief was that when you're looking for an example to mimic, you don't want to wade through someone else's project-specific details to get the idea. This article presents a collection of excellent ideas for identifying, disseminating, and evolving your software processes over time — a vital aspect of the Environment workflow.

With respect to standards and guidelines, if your organization does not mandate development guidelines applicable to the techniques and technologies that you intend to apply on your project, then you will either need to select or develop these guidelines as part of your Environment workflow efforts. The good news is that many development standards and guidelines are already available to your organization. For example, the Unified Modeling Language (UML) (Rumbaugh, Jacobson, Booch, 1999) provides an excellent basis for modeling notation standards, and coding guidelines for Java can be downloaded from `http://www.ambysoft.com/javaCodingStandards.html`. You don't need to reinvent the guidelines wheel. Instead, with a little bit of investigation, you often find that you can reuse existing standards and guidelines.

6.3 The Articles

6.3.1 "The Ten Commandments of Tool Selection" by Larry O'Brien
6.3.2 "Lessons Learned from Tool Adoption" by Karl Wiegers
6.3.3 "Timing is Everything" by Roland Racko
6.3.4 "Improve Your Process with Online 'Good Practices'" by Karl Wiegers

6.3.1 "The Ten Commandments of Tool Selection"

by Larry O'Brien

Start your personal tool-selection process with these time-tested axioms.

One of the basic tenets of Software Development magazine is that tools, without the right processes, are worthless, just as processes without the right tools are worthless. What's "right," we also feel, is extremely variable, depending on project size and whether a project is destined for shrink-wrap or internal deployment. It's especially dependent on the culture of the developers, their immediate managers, and their relationship to the rest of the organization. The subtleties of the relationships between people, products, and practices, are complex enough to fill the magazine month after month.

But while the subtleties are infinite, there are verities as well. Successful projects require a respect between developers and customers. Quality assurance must be as much a part of the design and construction process of a project as performance. You can find a great deal on these types of truths in such excellent books as *201 Principles of Software Development* by Alan M. Davis (McGraw-Hill, 1995) and the excellent series from Microsoft Press that includes *Code Complete* by Steve McConnell, *Dynamics of Software Development* by Jim McCarthy, and *Writing Solid Code* and *Debugging the Development Process*, both by Steve Maguire.

While a book could probably provide a thorough exploration of the task of tool selection (and wouldn't it be a fun one to write?), the problem goes remarkably undiscussed. To try to jump-start the discussion then, I'd like to offer what I think are the fundamental verities of tool selection. Commandments, if you will.

I. Software Will be Difficult to Create and Will Only Spring Forth from the Sweat of Your Brow

Anything worth automating is inherently complex. Understanding the features people would like automated is inherently complex. Creating something that is usable by people you haven't met is inherently complex. This has nothing to do with technology, or market segments, or hidden techniques known only to a select few. It's just the way it is.

Some tool vendors stay in business by selling the lie that they can make this truth disappear. The mere promise is so seductive that there are always a good number of people who will fork over their credit card numbers to anyone making the promise. Some people pay because they're gullible, others because they figure it's like a lottery and they might get lucky one of these years.

That's okay, but you must know early on if you're shopping for a tool or shopping for a lottery ticket.

II. When Shopping, Be Like Unto a Lion; When Shopping is Done, be Like Unto an Ass

Take the time to choose the right tool but once you've chosen it, commit to it. For any task there are a minimum of a dozen tools that *might* tackle it. Of those tools, half a dozen can tackle it. Of those, there are two or three that could be the right choice.

Not making the initial survey of tools broad enough is the number one tool selection mistake I see. People simply don't consider the right tools. "It's this type of problem, and this type of problem requires this type of tool." I've seen this (and, to be honest, fallen victim to it myself) time and time again. We all should know better than to assume that the weaknesses of a product or product type are permanent.

However, when the job of tool selection is done, for heaven's sake let it go. When the project hits its first big bump, it's natural to worry that you're wasting effort. But every project and every tool has a first, big bump. If you've done a good job selecting a tool, it's very, very unlikely that the latest release of a competing product is so radically better that jumping ships is worthwhile. Once you're ploughing your furrow, keep going straight ahead.

III. Honor thy Requirements Document

Comparative reviews generally have a list of criteria for comparing tools. If you're searching for a tool, you have a better set of criteria, your requirements. If you don't need Unicode support, you don't need it.

If you're reading for background, a reviewer's criteria are an excellent summary of leading-edge tool design in that category. Almost always, these lists combine the best features of all tools in the current generation, interweaving a few homilies like "ease of use" and maybe adding a recently ratified standards document or two. The very fact that, say, internationalization is mentioned as a criteria means that at least one tool in the current generation supports it. Reading reviews for such background information is a good way to honor the Second Commandment's injunction to keep up with the progress of those tools you don't think are capable.

IV. Do not Covet Thy Neighbor's Tool

For one thing, the grass is always greener on the other side of the fence. When someone is showing you a tool, it's natural to focus on the aspects that outperform your current tool. Maybe it has built-in support for multitasking and you're struggling with semaphores and blocking. Maybe it has a screen designer with built-in geometry management and you fill page after page with ratios of font height to button size to pane size. But rest assured, there are frustrations with every tool.

V. Render Unto Tool Makers That Which is Theirs

It's hard to admit that software tools can automate processes that you have always worked hard to do by hand. It's amazing how many people refuse to admit that compilers can generate better assembly code for a given piece of pseudocode than they can generate themselves. What was hard last year, much less the year before, much less what was hard all those years ago when you were first entering the profession, may now be routinely handled in a single function call.

The vast majority of tools out there are competent. They help you tackle tough jobs (perhaps not as well as you can imagine, but better than you can reasonably expect them to on any kind of a time budget.) There are occasions when it's appropriate to "roll your own," but they are few and far between. Most of the time, the roll-your-own mentality comes from two less-than-professional attitudes. One of these attitudes stems from the almost-certainly wrong idea that avoiding the purchase price and learning curve of a tool you buy will save money and time. The other attitude revolves around the selfish notion that being your own customer will make for a more relaxed, forgiving, and fun project.

VI. The Number of the Best Release of a Tool Shall be Three

The first release of a development tool is always rough, because the vendor has concentrated on functionality and not the interface. The second release is always dominated by fixing the errors of the first release. The fourth release is no good, and the fifth is right out, because of the checklist conundrum: after several years of competition and innovation, the review of any software category devolves to counting checkmarks against a feature list. Sadly, it's not even the percentage of checkmarks that's important, it's the absolute number. Most people use checklists not for item-by-item comparisons, but for pattern recognition, as in, "Which one of these tools has big gaps in functionality?"

VII. Tools Will be True to Themselves

A chainsaw makes one kind of sculpture, a chisel another. Software reflects the tools that built it. Before deciding on a tool, you should have an idea of how the tool's characteristics influence the delivered product. If that influence is unacceptable, you should seriously consider another tool, since fighting this type of inherent influence is time-consuming at best and futile at worst.

VIII. The Price of a Tool is Small, the Cost of a Tool is Great

The purchase price of a tool is nothing. You might make the common mistake of not valuing time spent getting up to speed on tools as much as you value "working time." But the time spent learning, integrating, and mastering a tool vastly outweighs the cost of any but the most expensive development tools.

There are at least two reasons why tool costs are much higher than they appear. One is related to the previous commandment about tools being true to themselves — tools will dictate the way you structure your working time with them and there's nothing you can do about it. If you're developing GUIs with a C++ compiler, your tool choice will dictate that you spend a lot of time waiting for the linking to be done. Another reason is cognitive dissonance, the mental quirk that puts a higher value on things you've already invested time in. No matter how well your intentions are regarding reevaluating tools when it comes to your next project, you're very likely to go with what you know.

This can lead to creeping obsolescence of your skills, which in our fast-moving industry proceeds at something more like a gallop, and nothing is costlier than losing your technical edge.

IX. Reviewers Lie with Lambs, You Will Dwell Among Wolves

Reviewers and computer journalists are usually treated a lot better than the average consumer. They don't pay for the software, they get special documentation, they get briefed on directions and strategies, they get special phone numbers for technical support, and the technical support they get is not from the summer interns staffing the phone lines. Despite everyone's best efforts to account for this preferential treatment, it's a fact that reviewers have a privileged relationship with vendors. This relationship cannot be viewed as representative of the partnership that should exist between you and your primary tool vendors.

A partnership is the appropriate model for a vendor and client relationship. Again, the previous two commandments dictate that a tool demands a lot from you — a certain way of doing things and a great investment in time. You have every right to voice your concerns to the vendor, and to expect not only an acknowledgment, but action. Once you invest in a tool, you have the right to expect the truth about delivery dates and capabilities. You should be able to ask questions like "When will this product run on Windows 95?" and get an answer with a technical, not marketing, basis.

And, finally, The Golden Rule of tool selection:

X. Recommend Unto Others That Which You Would Have Them Recommend Unto You

6.3.2 "Lessons Learned from Tool Adoption"

by Karl Wiegers

Software engineers talk a lot about using tools to help them perform development, project management, and quality tasks. They evaluate and buy a fair number of tools, but use few of them in their daily work. Selecting and installing a software tool is easy, but making effective tool usage integral to your group's software engineering practices is much harder.

Tool adoption initiatives often stumble because people don't fully appreciate the technical, managerial, and cultural factors that influence success. Here are eight lessons I've learned from groups attempting to use a variety of software development tools. Some of the lessons apply to any kind of tool adoption; others are most pertinent to computer-aided software engineering (CASE) tools for systems analysis and design. Keep these lessons in mind to increase the chance that your next tool investment will yield the desired payback.

Most tools that are available to the contemporary software developer or project manager belong to one of these categories:

- *Project management tools.* These help your team estimate, plan, and track schedules, resources, effort, and costs.
- *Analysis and design tools.* These help you document, analyze, and manage requirements or create design models.
- *Coding tools.* These include code generators and reformatters and code analysis and reverse engineering tools.
- *Quality improvement tools.* These include test planning and execution tools and static and run-time code analyzers.
- *Configuration management tools.* These let you track changes and defects, control access to files, and build your product from its components.

Most tools save time and reduce errors by automating a part of your development or management process. Tools that let you iterate on designs or accurately repeat tests can lead to higher quality and more successful projects. Defect-tracking tools ensure that problem reports do not get lost, and source control tools prevent one programmer's changes from obliterating another's. However, the first lesson to keep in mind is:

1. A tool is not a process.

For example, development groups sometimes believe that because they are using a tool to track defect reports, they have a defect-tracking process. In reality, a tool merely supports a process. The process defines the steps that you must perform to carry out some activity correctly. Team members must understand the steps, roles, and responsibilities of the process participants.

I once implemented a new change control system in a web development group. First, I wrote a change control process that the team members reviewed and accepted. We then beta-tested the new process using paper forms for a few weeks while I selected a commercial issue-tracking tool. Feedback from the beta helped us improve the process, and then I configured the tool to support the new process.

Introducing tools into a group can be disruptive. New ways of working demand new ways of thinking, and most people aren't eager to be forced out of their comfort zones. To succeed, the team members must agree on some conventions for using new methods and tools. This transformation requires both cultural and technical changes. This brings me to lesson two:

2. Expect the group to pass through a sequence of forming, storming, norming, and performing when it adopts new tools.

Forming, storming, norming, and performing describe the stages through which a new team progresses. These terms also characterize a software group trying to apply new approaches. During forming, the team selects a tool. During storming, team members debate how to use the tool, whether or not it is a good investment, and how to interpret the rules embodied in the tool or in its underlying methods. If you stop at storming, your team will not evolve to an improved future state that includes the tool. Storming is part of the learning curve, the cost of investing in new approaches that promise to yield long-term benefits.

Guide your team through the storming stage, relying on the team members' professionalism and willingness to compromise. It should then enter a norming stage, where the team agrees on how to apply the tool. Some individuals won't always get their way but everyone must be flexible to support the common good of improved team performance. The ultimate goal is performing, at which point the team achieves better results with the new tool than without it.

Any time someone is asked to change the way he or she works, you can expect to hear the question, "What's in it for me?" This is a normal human reaction; no one wants to go through an idle change exercise just because someone else had a cool idea. Sometimes, though, the change doesn't offer an immediate, obvious benefit for everyone. The lesson here is:

3. Ask "What's in it for us?," not "What's in it for me?"

Anyone attempting to introduce new software development methods or tools should be able to articulate the overall expected benefits to the project team, the organization, and the customers. For example, developers might not see the need to use code analyzer tools. After all, it might take them longer to complete their work if they have to run the code through a tool and fix any bugs it finds. However, explaining that it costs several times more to fix defects when the system testers find them than to use the tools during coding is a compelling argument. Quality practices that take one individual more time than his or her current work methods usually have a high return on investment downstream. This benefit might not be obvious to those who feel someone is making their jobs harder.

Tools are sometimes hyped as silver bullets for increasing productivity. Some tools do provide a substantial productivity benefit, as with automated testing tools that can run regression tests far faster and more accurately than a human tester can. However, the benefits from most tools come from improving the quality of the delivered product. Productivity gains come over the long term, as tools help you prevent defects and find existing bugs early and efficiently. Improved initial quality reduces late-stage, defect-repair costs. When rework is reduced, developers have more time available to create new products, rather than rehabilitate flawed ones. Keep lesson four in mind as you explore increased software development automation:

4. Tools affect quality directly, productivity indirectly.

I have heard development managers balk at acquiring development tools because they are expensive. "I can't afford a copy of Tool X for every team member at $2,000 per license," they protest. However, remember lesson five:

5. Weigh the price of the tool against the costs of not using it.

A good example here comes from requirements management tools. Licensing a powerful requirements management tool for a team of 10 software and quality engineers might cost $15,000. However, you don't have to go very far wrong on the project's requirements to waste $15,000 implementing unnecessary functionality .or re-implementing poorly under-stood and ill-communicated requirements. You cannot predict exactly how many errors such tools can help you avoid. Nonetheless, your tool acquisition decisions should consider the potential costs of not using a tool, in addition to the size of the check you'll have to write for it.

I worked with several small software development groups that used CASE tools over several years (see *Creating a Software Engineering Culture*, Dorset House, 1996). CASE tools let you draw models of requirements and designs according to a variety of standard modeling notations and languages. Commonly used models include data flow, entity-relationship, state-transition, class, sequence, and interaction diagrams. When they first came onto the scene in the 1980s, vendors claimed fabulous benefits from CASE tools. But some of the groups I have seen use CASE tools were simply documenting completed systems. This doesn't help you improve the software you build, but only helps you understand software you've already built. Lesson six is:

6. Use CASE tools to design, not just to document.

Use the tool's built-in methodology rules to validate diagrams and detect errors that are difficult for people to find. Your team should use the tools to iterate on design models before they start cutting code, because developers will never conceive the best design on their first attempt. Iterating on requirements and designs, rather than on code, is one way to improve quality and reduce product cycle times.

Our team had some energetic meetings in which we haggled over exactly what rules we should follow for various design models. Resolving these issues was critical to successfully implementing CASE in our group. We finally agreed on some conventions we could all live with. Lesson seven from our experience is:

7. Align the team on the spirit of the method and the tool, not on "The Rules."

Even if the developers do a great job on design, programs rarely match their designs exactly. You need to decide how to reconcile inconsistencies between design models and delivered software. If you want the CASE models to provide long-term benefits to the project, heed lesson eight:

8. Keep the information in the tool alive.

Ideally, you will update the models to match the reality of the system as it was built, and you'll keep the models current as you modify the system. This approach demands considerable effort and discipline. If the system evolves while the models remain static, inconsistencies between code and designs can cause confusion, wasted time, and errors during maintenance. If you decide not to update the design models, either discard them once they have served their purpose, or clearly identify them as representing the initial design, not the current software implementation.

Some CASE tools permit round-trip engineering — generating code from a design and regenerating design models automatically from source code. This approach guarantees the correctness of the as-built design documentation. You can repeat the model generation any time code changes are made to create a new set of accurate models.

Fitting Tools into Teams

A cultural transformation takes place as a team moves from manual methods to the discipline of using structured processes and tools. Facilitate this change by:

- Articulating why you selected the tools.
- Acquiring commitment from management.
- Selecting appropriate tools for your organization's culture and objectives.
- Training your team.
- Setting realistic expectations of the new technology's future benefits.

Some developers will balk at using new tools, maintaining that they can do better work with their current approach. For example, programmers who use interactive debuggers to perform unit testing (an inefficient process) may resist using test coverage tools. Some skeptics can be swayed when they see their teammates getting better results with the new tools, while others will never accept that there's a better way. Be alert for recalcitrant team members who sabotage the improvement effort by "proving" that the tools are worthless. Look for allies among the developer ranks, early adopters who are willing to adjust their current approaches to try new tools. Merge the tools into the way your team works, rather than reshaping the culture to fit the tool vendor's software development paradigm.

Educate your managers about the value of investing in tools, so they understand the tools aren't just toys to amuse technology-drunk software geeks. Obtain their commitment to spend the money you need to license the tools, train the team, and accept the short-term productivity hit from the learning curve. Ask management to set realistic and sensible expectations about how the team will incorporate tools into routine project activities.

The best chance for successful tool adoption comes if the tool capabilities will attack some source of pain your team is experiencing. For example, if developers spend countless hours on tedious searches for memory leaks or pointer problems, they might be willing to try run-time analyzers such as Rational's Purify or Compuware NuMega's BoundsChecker. Begin by identifying the areas where improvements are needed, then go to www.methods-tools.com and search for tools that might fit the bill.

Incorporating tools into a software organization requires more than dropping a user manual on every engineer's desk. Support the new tools with training in the underlying approaches, such as testing concepts or project estimation principles. Don't rely on tools to teach your developers the fundamental methods, any more than you can learn arithmetic by using a calculator. The individuals who I've seen use tools most effectively have a solid understanding of the methods and principles their tools implement.

As you pursue effective software development and management tool use, respect the learning curve, which will likely make your first attempts to use new tools actually take longer than your previous methods did. Share your successes and failures with the team, so everyone can learn from the experiences of others. Work toward a long-term objective of providing each of your developers with a robust, integrated automation suite that will increase their productivity and the quality of the software they deliver.

6.3.3 "Timing is Everything"

by Roland Racko

Building your first object-oriented, component-based, client/server distributed intranet buzzword-compliant project? There are many things to buy for such a project. But in which order should you choose the methodology, tools, language, integrated development environment, compiler vendor, database, external libraries, documentation standards, reuse standards, and hardware you need?

Having worked on a number of these startup projects, I've found the following points guide me well. They keep the client happy and the project on time. First, compulsory single-vendor strategies only work for a short while, namely until the vendor gets too stodgy and begins to put more effort into creating fear, uncertainty, and doubt in its customers than in creating innovative solutions for them. Single-vendor purchase strategies can also leave you in the dust in these days of small garage shop innovation that turns the technological world upside down.

Buying "best of breed" across several vendors is also risky, but not for the usual reason: integration problems between various products. The problem with best of breed is that arriving at the criteria for "best" is typically done by reading magazine reviews or appointing a committee to study the products in the abstract using a generalized rating scheme. While it used to be reasonable to try to set a company-wide standard when the product marketplace was less volatile, today a one-time judgement of the marketplace becomes outdated in six months.

More important, such a decision does not account for the needs of each business problem. An old rule of thumb, not yet repealed by any ANSI committee, states that a system will go together quickly and have a low maintenance cost to the extent that the shape of its technical solution mirrors the shape of the business problem. Both compulsory single-vendor purchases and generalized best-of-breed approaches are one-size-fits-all ideas that tend to force-fit a solution shape. The consequences of that constraint are subtlely degraded implementation and maintenance efforts with inherent difficulties that become hard to trace back to their real cause.

Diving Into the Tar Pit

From these observations comes my first recommendation — don't make any choices without examining the business problem first. Deciding which tool to use for the examination creates a bit of a chicken and egg problem. The tool you use can force its own shape onto the problem, potentially moving the solution away from the optimum. Yet, if the problem is sufficiently complex, you'll need some tools just to lay out some rough ideas. Use something simple, perhaps narrative English and a diagramming tool like Visio. Your basic task at this stage is to determine how the system is rich and to what extent performance is an issue.

By rich, I mean discovering what the preponderant problem area in the system is. Is it the GUI? The number of objects? The flexibility of the objects? The linking complexity of the probable web pages? The complexity of the database? The complexity of the business logic? The sequence of events? Another way of thinking about it is to examine the system, using any tool at all, to determine what the top few technical programming issues are: what is going to keep you awake at night? Let's call these central technical issues the project tar pit. Add to that project tar pit any preliminary concerns about run-time performance, whether it is inter-object communication bandwidth, number of users, database access time, ease of distributing updates to business rules, and so on.

Applying Insights

When you can state your central technical issues on one page, you are ready to make your first choice: methodology (process). The methodology you choose should be the one that helps you best comprehend the project tar pit you have outlined. All methodologies play favorites, so some will be better at one problem than others. Once you find the methodology that seems to have had the most success with your type of tar pit, resist the temptation to make it a company standard. Don't waste your time on standardizing at this point. Any current methodology will be replaced by new and better stuff in a few years. Some garage shop will see to it.

A possible snag is that some methodologies are tied to proprietary analysis and design tools. Avoid this kind of early constraint unless there are very compelling reasons not to. Instead, pick a more open methodology, even if it is a lesser choice, so you can let the business problem itself be the stronger factor in the remaining choices.

Now your choices get a little more complex. Two global factors, unrelated to the business problem, strongly modify the sequence in which things are chosen for an object-oriented project. The first of these is ramp-up time. For example, at some point you are going to have to choose an integrated development environment and coding language. Learning those will take time, so ideally that learning should begin before you start coding. But at the moment, you do not have enough details about the problem to make such a decision. Therefore, you must use the methodology to do some analysis and possibly some design so the language and integrated development environment choices that might be suitable become clearer. Then start a parallel effort at trying them out so by the time you get to coding, you will be able to make a rational decision based both on problem shape and personal experience.

Other factors may come into force during this parallel effort. Suppose that reusability were a strong emphasis on this project. That might strongly weight your first choice in favor of, say, Eiffel. But the web nature of the project might argue in favor of Java. It does not follow that either has the most comfortable development environment at the moment.

My recommendation would be to carry on the parallel effort of testing several integrated development environments and languages as analysis and design proceeds. The project tar pit may change during this period, or you may refine your ideas about it. When it comes time to code, factor in all the project priorities and then pick the language that helps the most with whatever remains of the tar pit. If language is no longer a factor, then pick the most comfortable development environment and let the language come from that. It is O.K. that there might be more than one language and development environment. There could easily be four of each in our buzzword-compliant sample — C# (C-sharp), HTML, Java, and SQL, not to mention some vendor's proprietary scripting language.

Database choice and performance issues have similar timing. You need to use the chosen methodology to do enough spade work to determine where potential performance bottlenecks might occur. It's important to determine what sort of database architecture will work best. All of this may be very intertwined, with database choices interacting with hardware choices interacting with external library choices interacting with network loading.... Well, I didn't say it would be a piece of object-oriented cake.

Unless other factors are a consideration, you should choose hardware last. It should be chosen when analysis is more or less finished and design is nearly complete. That way, the programming solution determines the most effective hardware. Other factors are the typical gotcha in this ideal sequence — politics, purchasing lead time, the need for a box to do prototyping, the need for a box to test out the language and development environment, the need to do performance simulation, and so on.

Proceed with Caution

The second global factor that influences the sequence of choosing products is related to the shop environment rather than the business problem. The Racko standard change coefficient is 2 + or − 1, meaning that a company can only absorb a few changes at a time.

If you are changing methodology, language, development environment, hardware, database, and networking style all at once, you are exceeding this coefficient by a considerable margin. Potentially, none of them could come off very well or all of them could take much more time than anybody guessed. So a group might decide to avoid making a choice, at least just yet, and stick with the current in-house facility, even if that means a less-than-optimum solution. For example, a group might decide not to use any new languages, staying instead with its current shop languages and kludging things up a bit or reducing the specification functionality just to keep the rate of change in the comfort zone. The possibility of a new language would then be reexamined in the next object-oriented web project.

Documentation and reuse standards are not influenced by the shape of a single business problem or solution and can be chosen independently of project schedule. It would be nice to have some preliminary standards before you begin, however. If this is your first object-oriented project, however, it is probably not the right time to deal with standards about reuse. Better to let people absorb all the other changes, get the product out the door, and then mine it for its reuse potential, deriving reuse standards during that mining effort.

Oh yes, one last thing — don't forget to obtain good upper management sponsorship.

6.3.4 "Improve Your Process with Online 'Good Practices'"

by Karl Wiegers

Most software developers are allergic to paper. As organizations improve their software development and management processes, though, they typically find they need some additional documentation. More structured development life cycles — the result of process improvement — may present developers with new expectations about activities to perform and deliverables to create. An important part of successful process improvement is to develop a supporting infrastructure of process and project documentation. This documentation captures the collective experiences, insights, and wisdom — the good practices — of developers and managers throughout the community. An intranet is a perfect way to share that accumulated knowledge.

Documentation: What and Why

People who are asked to apply new practices will benefit from written guidelines. Documentation may include process descriptions, procedures, checklists, and other work aids. These items help project team members understand expectations embedded in improved approaches. They also provide a mechanism for sharing the best known ways to perform key activities like gathering requirements, estimating project effort and duration, and controlling changes to the product.

The contents and structure of a "good" document are not intuitively obvious. A collection of document templates and examples will help each new project team create its key documents more efficiently. If you adopt common templates, similar work products developed on multiple projects will have similar contents and structure. Each new project doesn't have to invent a requirements specification or configuration management plan from scratch, perhaps omitting important sections along the way. Good examples of key work products collected from previous projects give developers a model for their own project's documents, and may also save them time through reuse of common text.

In this article, I'll describe how one company's repository of useful software engineering project and process documents helped its software teams share information and improve development process. This resource, which I helped develop, is now being used by various software project teams, thereby leveraging the investment made in developing the resource. I'll also describe the contents and structure of the document collection, as well as the way our "good practices" project team organized and ran the project that created the collection.

Motivation and Philosophy

The company's process improvement leaders felt that a company-wide resource of key documents could assist all software project teams. Many departments had already launched process improvement activities, and several had begun building their own collections of example work products and templates. They'd already assembled a large, central collection of examples, but it was underused for several reasons. It existed only in hard copy form, which made it difficult to share the contents across the company or search for specific items. The contents varied in quality, since no filtering or revision was done prior to including submitted items.

Additionally, the contents addressed the needs of only a portion of the company's development projects.

From a practical point of view, you should invent new procedures only as a last resort; instead, look to adopt or adapt existing procedures to meet your needs. For example, our project team knew that several organizations were using the Software Capability Maturity Model (CMM) to guide their process improvement activities, and that they needed to create a number of documented procedures described in the CMM. We wanted to leverage those procedures across our company to make our process improvement activities more efficient.

To take advantage of the available resources, we built a corporate-wide intranet-based collection of software engineering "good practices." This central resource would reduce the effort individual departments needed to create software development procedures, because they would have good examples with which to begin. The collection also would house good examples of key plans and other work products created on previous corporate projects. Following a pilot project to explore the concept and evaluate approaches, I was invited to lead the project to implement the repository.

We called our site the Good Practices web site. We specifically avoided the term "best practices" for several reasons. First, it begs the question of who determines what is best and what the reference points are. Second, we expected the initial collection to be improved upon over time, as project teams gained experience developing and applying new processes and created better examples. Thus, we focused on building a collection of useful, but not necessarily perfect, artifacts that would be better than the resources presently available.

We assembled a small, part-time team to carry out this project; as project leader, I was committed nearly full-time. Participants included members of a software engineering process group that supported a large product development division, a webmaster, and other individuals who had a strong interest in quality and process. Our initial strategy was to:

- Identify the appropriate contents of the collection.
- Devise a scheme for organizing and labeling the many documents we expected to encounter.
- Assemble small volunteer teams of subject matter experts of various software practice domains to help evaluate, revise, and create suitable documents.
- Devise a suitable web architecture for making the documents as accessible as possible.
- Write a detailed, realistic project plan.
- Follow the plan to create and populate the repository, adjusting the plan as needed.

A fundamental philosophy of the project was that "less is more," in two respects. First, we wanted to include just a few documents of each type. If you have a dozen sample requirements specifications from which to choose, you're likely to create your own, rather than study all 12 and pick the format most suitable for your needs. Therefore, we included just a few templates and examples of each key document type. These represented approaches that were suitable for different projects, such as a small information system and a large embedded systems product.

The second "less is more" notion led us to trim actual project documents down to a manageable size that simply illustrated the structure, contents, and style of the document. If you're looking for an example to mimic, you don't want to wade through someone else's project-specific details to get the idea. We took considerable editorial license to modify candidate document examples to meet our goals of size, detail, and layout.

Our Approach

We began by assembling a shopping list of items we thought would be valuable to include. The list combined items from the CMM, activities and deliverables defined by the product development life cycle followed by many projects, and solid software engineering approaches. The project team also identified candidate items we might harvest from the existing local collections throughout the company.

The initial shopping list identified well beyond 200 different policies, process descriptions, procedures, work aids, checklists, templates, and work product examples. Some of these were more important than others, so we prioritized the list into three levels. Amazingly, about one-third of the items ended up in each priority level, which made it much easier to focus on doing the important things first. We realized that the lowest priority items would probably never be acquired, since we did not have infinite resources for this project. The initial version of the web site included 90% of the high priority documents, 44% of the mediums, and only 20% of the lows, demonstrating that we did a good job of investing our effort where it could yield the greatest benefit.

The project team also applied sound requirements engineering practices to this project. We began by brainstorming a list of use cases, each describing one objective a prospective user of the Good Practices web site might try to achieve. We have found the use case approach to be valuable for other web development projects, as they help the project team focus on things a user could do when visiting the site. Some of these use cases were:

- Find a software project plan template that's suitable for use on a large embedded software project and compliant with the pertinent IEEE standard.
- Find a procedure for managing requirements that complies with the expectations described in the CMM.
- Download forms to use for conducting a code inspection.
- See what training classes are available for a specific subdomain of software engineering.

While most of the use cases we identified were addressed by this project, we concluded that others were out of scope. We addressed some of those by providing hyperlinks from the Good Practices web site to other sites (such as our training catalog), while others we simply didn't address. To control scope creep, it helps to carefully define your project's scope in the first place, and then evaluate whether each proposed requirement, use case, or function fits within that scope.

We also modeled the web site's data architecture using an entity-relationship diagram and data dictionary. We modeled the user interface architecture using a dialog map, which shows the various proposed web pages and navigation pathways among them. Finally, we developed a series of evolutionary web site prototypes, which were evaluated by our subject matter experts and eventually became the product. All of these requirements and design practices helped us develop a usable and robust product that was well received by our user community of more than 1,000 developers and managers.

The Good Practices project team delegated much of the evaluation and improvement (and even creation) of suitable documents to the subject matter expert teams. The subject matter experts came from the development, management, process improvement, and quality engineering ranks. Without the active participation of these volunteers, we would probably still be working on the project.

Good Practices Contents

We organized the contents of the Good Practices collection so users could easily locate what they needed. We developed three searchable catalogs that listed all of the site's contents. One catalog was organized by CMM key process areas, since many departments in the company were using the CMM to guide their process improvement activities. Another catalog was based on the activities and deliverables defined for each phase of the product development life cycle being followed by several hundred developers in various organizations. We identified available procedures and checklists for the activities, and available templates and examples for the deliverables.

The third catalog was organized by software engineering practice areas. Table 6.1 shows the hierarchy of practice areas we identified. Every document in the collection was uniquely identified by a label that consisted of the practice area to which it belonged, a document type code (for example, work product example, procedure, and template), and a sequence number.

By scouring the company, we found good examples of most document categories. Many of the best examples came from projects or organizations that already had a well-established software process improvement program, thereby increasing the return on investment from their improvement initiative.

To plug some holes in the collection, we selected items from a commercial source, EssentialSET from the Software Productivity Center (www.spc.ca). EssentialSET includes more than 50 document examples and templates from a hypothetical medium-sized project, covering the majority of the software engineering practice areas we defined. This is a good way to jump-start your efforts to build a Good Practices resource.

The publication process for preparing documents for installation on the intranet was quite elaborate. We decided to create a master copy of each document in Microsoft Word. We converted documents intended to be read-only to Adobe Portable Document Format (PDF). We published templates intended to be downloaded and adapted on the web site in both Microsoft Word and FrameMaker format. We wrote a detailed procedure for the many steps involved in the publication process, which might include OCR scanning of hard copy documents, adding descriptions, editing for size or confidentiality, and reformatting.

Tracking the status of this project was not trivial. We used a series of linked Microsoft Excel worksheets to monitor the evaluation and publication status of about 300 separate items. For accepted candidates, we tracked these key milestone dates: accepted for inclusion; permission received from the owner (to include item on web site); converted to Microsoft Word; all edits complete; converted to delivery format; delivered to the webmaster; installed; and tested.

Each month we tracked the number of documents that were delivered in each practice area and priority level. Additionally, we tracked the time that all project participants, including subject matter experts, spent on the project every month, and the time spent on the various document conversion operations. These metrics provided a complete and accurate picture of the project's cost, achievements, and status at all times.

Web Architecture

In the web page architecture we implemented, each practice area from Table 6.1 has a Practice Area page. This describes the practice area in general, and provides information about, or links to, available training classes, tools, pertinent books, local subject matter experts, and other useful resources.

Each Practice Area page lists the document categories available for that practice area. For example, the Requirements Management practice area includes document categories for software requirements specification, interface specification, requirements management procedure, and others. Each document category page provides links to several specific documents or templates, each of which has a short description to help the user select examples that are most pertinent to his or her needs and project type.

Table 6.1   Software engineering and management practice areas.

Practice Area	Subpractice Areas
Design	Architecture Design Detailed Design User Interface Design
Implementation	Coding Integration
Maintenance	(none)
Project Management	Estimation Project Planning Project Tracking Risk Management
Requirements Engineering and Management	Requirements Gathering Requirements Management
Software Configuration Management	Configuration Auditing Configuration Control Configuration Identification Configuration Status Accounting
Software Quality Assurance	Auditing Metrics Peer Reviews
Software Subcontract Management	Accepting Subcontracted Materials Proposals and Contracting Tracking Subcontract Activities
Testing	Beta Testing Integration Testing System Testing Unit Testing

Lessons Learned

At the conclusion of the project, the team captured and documented the lessons we learned. Many lessons came out of things we did right, though others were learned because we missed the best approach the first time around. These lessons fell into three major categories:

1. Doing It Right. Develop use cases to focus scope; develop multiple prototypes; provide prototype evaluation scripts to the users; and document procedures for publishing documents on the web site.

2. Controlling the Project. Plan the project thoroughly; build a shopping list of documents needed; set and respect three levels of work priority; track document and project status rigorously; review the risk list periodically; walk through the project work breakdown structure as a team to spot missing tasks; and record action items and decisions made at meetings.

3. Stability Over Time. Make sure webmasters inform us when links change; test all external links periodically as part of the project maintenance plan; and review and improve contents periodically to keep them current and useful.

Process improvement projects that rely on the volunteer contributions of participating co-workers face a major risk: this commitment may not float to the top of a busy developer's priority list. If a subject matter expert's team was not getting the job done, we had no choice but to replace the leader or other members with people who would contribute effectively. We also decreed that the decision-making quorum for any meeting would be whoever showed up, thereby keeping the project rolling along.

This project met most of its major milestones, and it was delivered on schedule and under budget. During the past six months, hundreds of developers, quality professionals, and process improvement leaders downloaded more than 3,500 documents, suggesting that software teams find this to be a valuable resource.

It's difficult to quantify the return on investment for such a project. However, this consolidated resource of software engineering good practices clearly made the process and project improvement activities at this company more efficient. Look around for the most effective ways your project team members are performing key tasks. Then, take the time to capture their wisdom and share it with the rest of the organization through your own "good practices" collection.

7

Parting Words

We have known the fundamentals of the software process for years. One has only to read classic texts such as Fred Brooks' *The Mythical Man Month* (1995), originally published in the mid-1970s, to see that this is true. Unfortunately, as an industry, we have generally ignored these fundamentals in favor of flashy new technologies promising to do away with all our complexities, resulting in a consistent failure rate of roughly 85% on large-scale, mission-critical projects. Seven out of eight projects fail — that is the cold, hard truth. Additionally, this failure rate, and embarrassments such as the Y2K crisis, are clear signs that we need to change our ways. It is time for organizations to choose to be successful, to follow techniques and approaches proven to work in practice, and to follow a mature software process.

A failure rate of roughly 85% implies a success rate of only 15%.
Think about it.

The Inception phase is where you define the project scope and the business case — the justification, for your system. In a nutshell, the goals of the Inception phase are to:

- describe the initial requirements for your system,
- determine the scope of your system,
- identify the people, organizations, and external systems that will interact with your system,
- develop an initial risk assessment, schedule, and estimate for your system,
- justify both the system itself and your approach to developing/obtaining it, and
- develop an initial tailoring of the Unified Process to meet your exact needs.

Your efforts during the Inception phase should create and/or evolve a wide variety of artifacts, including:

- a vision document,
- an initial requirements model (10–20% complete),
- an initial project glossary,
- a business case,
- an initial domain model (optional),
- an initial business model (optional),
- a development case (optional) describing your project's tailored software process, and
- an architectural prototype (optional).

7.1 Looking Towards Elaboration

The Elaboration phase focuses on detailed analysis of the problem domain and the definition of an architectural foundation for your project. To move into the Elaboration phase, you must pass the Lifecycle Objective (LCO) milestone (Kruchten, 2000). To do so, during the Inception phase, you must achieve:

- Consensus between project stakeholders as to the project's scope and resource requirements
- An initial understanding of the overall, high-level requirements for the system
- A justification for your system that includes economic, technological, and operational issues
- A credible, coarse-grained schedule for your entire project
- A credible, fine-grained schedule for the initial iterations of the Elaboration phase
- A credible risk assessment and resource estimate/plan for your project
- A credible initial tailoring of your software process
- A comparison of your actual vs. planned expenditures to date for your project
- The optional development of an initial architectural prototype for your system

---

During the Inception phase, you define the project scope and the business case. During the Elaboration phase, you will define the architectural foundation for your system.

Software development, operations, and support are complex endeavors, ones that require good people, good tools, good architectures, and good processes to be successful. This four-volume book series presents a collection of best practices for the enhanced lifecycle of the Unified Process, best practices published in *Software Development* (www.sdmagazine.com), and written by luminaries of the information industry. The adoption of the practices best suited to your organization is a significant step towards improving your organization's software productivity. Now is the time to learn from our past experiences; now is the time to choose to succeed.

Scott's Really Bad Zen Poetry

A tradition of this book series is to inflict a really bad poem on you at the very end of the book, potentially making you regret having purchased the book in the first place. Does this make any sense[1]? Likely not, but we're doing it anyway.

The Zen of Inception

Project stakeholders jostle for position,
    wielding competing visions like swords.
Will rampant politics yield to consensus?
Melting snowball extinguishes flame.

---

1. Actually, now Scott can claim to be a published poet, giving him an incredibly good pick-up line whenever he is in a bar. There is a method to his madness.

Appendix A

Bibliography

Ambler, S.W. 1995. Reduce Development Costs with Use-Case Scenario Testing. *Software Development*, July.

Ambler, S.W. 1998. How the UML Models Fit Together. *Software Development*, March.

Ambler, S.W. 1999. Debunking OO Myths. *Software Development*, February.

Ambler, S.W. 2000. Requirements Patterns. *Software Development*, May.

Beyer, H. 1997. Data-Based Design. *Software Development*, January.

Bianco, N. 1999. A Business Case for QA and Testing. *Software Development*, February.

Comaford, C. 1995. What's Wrong with Software Development. *Software Development*, November.

Constantine, L. 1998. Real-Life Requirements. *Software Development*, May.

Constantine, L. 1998. Scaling Up Management. *Software Development*, November.

Constantine, L. 1999. Habits of Productive Problem Solvers. *Software Development*, August.

Douglass, B.P. 2000. Organizing Models the Right Way. *Software Development*, August.

Downs, A. 2000. From Engineer To Technical Lead. *Software Development*, April.

Geier, J. 1996. Don't Get Mad, Get JAD. *Software Development*, March.

Glassett, S. 1999. Effective Resource Management. *Software Development*, September.

Goldsmith, R. 1999. Plan Your Testing. *Software Development*, April.

Gottesdiener, E. 1999. Decoding Business Needs. *Software Development*, December.

Gottesdiener, E. 2000. Capturing Business Rules. *Software Development*, January.

Highsmith, J. 1996. Mission Possible. *Software Development*, July.

Hohmann, L. 1998. Getting Started with Patterns. *Software Development*, February.

Keuffel, W. 1995. Don't Fence Me In. *Software Development*, February.

Launi, J.D. 1999. Creating a Project Plan. *Software Development*, May.

Lockwood, L.A. 1999. Learning the Laws of Usability. *Software Development*, October.

Margulies, B.I. 2000. Your Passport to Proper Internationalization. *Software Development*, May.

McConnell, S. 1997. Managing Outsourcing Projects. *Software Development*, December.

Minkiewicz, A. 2000. Software Measurement: What's in It for Me?. *Software Development*, March.

O'Brien, L. 1995. The Ten Commandments of Tool Selection. *Software Development*, November.

Racko, R. 1996. Timing is Everything. *Software Development*, October.

Roetzheim, W. 2000. Estimating Internet Development. *Software Development*, August.

Rothman, J. 1999. Determining Your Project's Quality Requirements. *Software Development*, February.

Runge, L. 1995. Thirteen Steps to a Successful System Demo. *Software Development*, March.

Thomas, D. 1998. Web Time Software Development. *Software Development*, October.

Whitten, N. 1995. Selecting the Best Vendor. *Software Development*, November.

Wiegers, K. 1997. Metrics: 10 Traps to Avoid. *Software Development*, October.

Wiegers, K. 1998. The Seven Deadly Sins of Software Reviews. *Software Development*, March.

Wiegers, K. 1998. A Project Management Primer. *Software Development*, June.

Wiegers, K. 1998. Know Your Enemy: Software Risk Management. *Software Development*, October.

Wiegers, K. 1998. Improve Your Processes with Online 'Good Practices.' *Software Development*, December.

Wiegers, K. 1999. Lessons Learned from Tool Adoption. *Software Development*, October.

Wiegers, K. 1999. A Software Metrics Primer. *Software Development*, July.

Wiegers, K. 1999. Customer Rights and Responsibilities. *Software Development*, December.

Wilkinson, N. 1995. CRC Cards for Analysis. *Software Development*, October.

B

Contributing Authors

Beyer, Hugh   Hugh Beyer has worked in the industry as a programmer, designer, and system architect since 1986 and is a co-founder of InContext Enterprises Inc. — a firm specializing in process and product design consulting.

Bianco, Nicole   Nicole Bianco has worked in software development and management for more than 30 years. She is the software initiatives leader in the Internet and networking group at Motorola.

Comaford, Christine   Christine Comaford is a columnist for *PC Week*. Her website is www.christine.com.

Douglass, Bruce Powel   Bruce Powel Douglass is a frequent contributor to *Software Development* and is the author of *Doing Hard Time: Developing Real-Time Systems with UML, Objects, Frameworks and Patterns* (Addison-Wesley, 1999).

Geier, Jim   Jim Geier is a senior systems consultant at TASC Inc. in Dayton, Ohio, where he provides information system engineering consultation to commercial companies and government organizations.

Glassett, Susan   Susan Glassett is vice president of development at ABT Corp.

Goldsmith, Robin   Robin Goldsmith is president of Go Pro Management Inc. in Needham, Massachusetts and can be reached through their website, www.gopromanagement.com.

Gottesdiener, Ellen   Ellen Gottesdiener is a facilitator, trainer, and consultant in Carmel, Indiana. You can reach her through her website, www.ebgconsulting.com.

Highsmith III, James A.   James (Jim) A. Highsmith III began his career developing software for the Apollo spacecraft program, is a principal at Information Architects Inc., and isthe author of *Adaptive Software Development* (Dorset House Publishing, 2000).

Hohmann, Luke Luke Hohmann is an experienced software professional whose career has included several successful startups. He was most recently VP of Engineering and Product Development for Aurigin Systems, Inc. and is the author of *Journey of the Software Professional: A Sociology of Software Development* (Prentice Hall, 1996), as well as numerous articles on software development. He is presently working on starting yet another software company in Silicon Valley.

Holtzblatt, Karen Karen Holtzblatt has designed products and processes in the computer industry since 1989, including the Contextual Inquiry approach to gathering field data. She is a co-founder of InContext Enterprises Inc. — a firm specializing in process and product design consulting.

Keuffel, Warren Warren Keuffel writes on a variety of software engineering topics from his home in a Salt Lake City, Utah suburb.

Launi, Joseph D. Joseph D. Launi is a director in the project management division of Spectrum Technology Group Inc.

Lockwood, Lucy Lucy Lockwood is the co-author with Larry Constantine of *Software for Use* (Addison-Wesley, 1999). You can reach her through `www.foruse.com`.

Margulies, Benson I. Benson I. Margulies is vice president and chief technology officer for Basis Technology Corp. — an internationalization services firm in Cambridge, Massachusetts. His experience ranges from secure operating systems to object-oriented databases to cable TV set-top box applications.

McConnell, Steve Steve McConnell is the author of *Code Complete* (Microsoft Press, 1993), Jolt-award winning *Rapid Development* (Microsoft Press, 1996), *Software Project Survival Guide* (Microsoft Press, 1997), numerous technical articles, and is the editor of *IEEE Software*.

Minkiewicz, Arlene Arlene Minkiewicz is chief scientist at PRICE Systems L.L.C. She speaks frequently on software measurement and estimating and has published articles in the *British Software Review*.

Nix, Lynn Lynn Nix is a consultant specializing in project management, accelerated software development, and software process improvement.

O'Brien, Larry Larry O'Brien is the president of an Internet startup firm and a contributing editor of *Software Development* magazine.

Racko, Roland Roland Racko is a veteran consultant usually found functioning as a troubleshooter on large client/server projects.

Roetzheim, William William H. Roetzheim is founder and CEO of the Cost Xpert Group in Jamul, California. An authority on software cost estimating, he has written 16 technical books and more than 100 articles. Visit his company website at `www.costXpert.com`.

Rothman, Johanna Johanna Rothman provides consulting and training to improve product development practices. She has more than 20 years experience in the software industry.

Runge, Larry Larry Runge is the vice president and chief information officer at Wheels Inc. Currently, he is leading teams that develop large-scale client/server information systems.

Thomas, Dave Dave Thomas is the founder of Object Technology International Inc. and has more than 30 years experience as a developer, professor, architect, product manager, director, president, and CEO.

Weber Morales, Alexandra   Alexandra Weber Morales is Editor in Chief of *Software Development* magazine. After attending Bryn Mawr College in Pennsylvania, she spent several years as a freelance writer, primarily covering healthcare. She joined Miller Freeman, Inc. (now CMP Media, Inc.) in January 1996 and spent three years traveling Latin America as chief editor of a Spanish- and Portuguese-language medical imaging technology magazine. She speaks fluent Spanish, French, and Portuguese in addition to her native English. Her interests once included mid-century internal combustion engines (she rebuilt several Volkswagens and one 1964 Chevrolet P-10), but now tend toward graphic design, music, and public transportation.

Wiegers, Karl   Karl Wiegers is the principal consultant at Process Impact, the author of the Jolt Productivity Award-winning *Creating a Software Engineering Culture* (Dorset House, 1996), and *Software Requirements* (Microsoft Press, 1999). Karl Wiegers is also a contributing editor to *Software Development*.

Wilkinson, Nancy   Nancy Wilkinson is part of the Software Development Technology Group at AT&T Bell Laboratories and is the author of *Using CRC Cards* (SIGS Books, 1995).

Whitten, Neil   Neil Whitten, president of the Neal Whitten Group, is a speaker, trainer, consultant, and author specializing in software project management. He provides periodic project reviews to identify ways to improve project efficiency and is the author of *Managing Software Development Projects: Formula for Success, Second Edition* (John Wiley & Sons, 1995).

C

References and Recommended Reading

Printed Resources

Ambler, S.W. 1998a. *Building Object Applications That Work: Your Step-By-Step Handbook for Developing Robust Systems with Object Technology.* New York: SIGS Books/Cambridge University Press.

Ambler, S.W. 1998b. *Process Patterns — Building Large-Scale Systems Using Object Technology.* New York: SIGS Books/Cambridge University Press.

Ambler, S.W. 1999. *More Process Patterns — Delivering Large-Scale Systems Using Object Technology.* New York: SIGS Books/Cambridge University Press.

Ambler, S.W. 2001. *The Object Primer, 2nd Edition: The Application Developer's Guide to Object Orientation.* New York: SIGS Books/Cambridge University Press.

Ambler, S.W. and Constantine L.L. 2000a. *The Unified Process Construction Phase.* Lawrence, KS: CMP Books.

Ambler, S.W. and Constantine L.L. 2000b. *The Unified Process Elaboration Phase.* Lawrence, KS: R&D Books.

Ambler, S.W. and Constantine L.L. 2001. *The Unified Process Transition Phase.* Lawrence, KS: CMP Books.

Bassett, P.G. 1997. *Framing Software Reuse: Lessons from the Real World.* Upper Saddle River, NJ: Prentice Hall, Inc.

Baudoin, C. and Hollowell, G. 1996. *Realizing the Object-Oriented Life Cycle.* Upper Saddle River, NJ: Prentice Hall, Inc.

Beck, K. and Cunningham, W. 1989. "A Laboratory for Teaching Object-Oriented Thinking." *Proceedings of OOPSLA'89.* pp. 1–6.

Beck, K. 2000. *Extreme Programming Explained — Embrace Change.* Reading, MA: Addison Wesley Longman, Inc.

Bennett, D. 1997. *Designing Hard Software: The Essential Tasks.* Greenwich, CT: Manning Publications Co.

Binder, R. 1999. *Testing Object-Oriented Systems: Models, Patterns, and Tools.* Reading, MA: Addison Wesley Longman, Inc.

Booch, G. 1996. *Object Solutions — Managing the Object-Oriented Project.* Menlo Park, CA: Addison Wesley Longman, Inc.

Booch, G., Rumbaugh, J., and Jacobson, I. 1999. *The Unified Modeling Language User Guide.* Reading, MA: Addison Wesley Longman, Inc.

Brooks, F.P. 1995. *The Mythical Man Month.* Reading, MA: Addison Wesley Longman, Inc.

Brown, W.J., McCormick III, H.W., and Thomas, S.W. 2000. *AntiPatterns in Project Management.* New York: John Wiley & Sons Ltd.

Buschmann, F., Meunier, R., Rohnert, H., Sommerlad, P., and Stal, M. 1996. *A Systems of Patterns: Pattern-Oriented Software Architecture.* New York: John Wiley & Sons Ltd.

Champy, J. 1995. *Reengineering Management: The Mandate for New Leadership.* New York: Harper Collins Publishers Inc.

Chidamber, S.R. and Kemerer C.F. 1991. "Towards a Suite of Metrics for Object-Oriented Design." *OOPSLA'91 Conference Proceedings.* Reading, MA: Addison Wesley Longman, Inc., pp. 197–211.

Coad, P. and Mayfield, M. 1997. *Java Design: Building Better Apps and Applets.* Englewood Cliff, NJ: Prentice Hall, Inc.

Compton, S.B. and Conner, G.R. 1994. *Configuration Management for Software*. New York: Van Nostrand Reinhold.

Constantine, L.L. 1995. *Constantine on Peopleware*. Englewood Cliffs, NJ: Yourdon Press.

Constantine, L.L. & Lockwood, L.A.D. 1999. *Software for Use: A Practical Guide to the Models and Methods of Usage-Centered Design*. New York: ACM Press.

Constantine, L.L. 2000a. *The Peopleware Papers*. Englewood Cliffs, NJ: Yourdon Press.

Constantine, L.L. 2000b. *Managing Chaos: The Expert Edge in Software Development*. Reading, MA: Addison Wesley Longman, Inc.

Coplien, J.O. 1995. "A Generative Development-Process Pattern Language," *Pattern Languages of Program Design*. Reading, MA: Addison Wesley Longman, Inc., pp. 183–237.

DeLano, D.E. and Rising, L. 1998. "Patterns for System Testing," *Pattern Languages of Program Design 3*, Martin, R.C., Riehle, D., and Buschmann, F., (eds.). Reading, MA: Addison Wesley Longman, Inc., pp. 503–525.

DeMarco, T. 1997. *The Deadline: A Novel About Project Management*. New York: Dorset House Publishing.

Douglass, B.P. 1999. *Doing Hard Time: Developing Real-Time Systems with UML, Objects, Frameworks, and Patterns*. Reading, MA: Addison Wesley Longman, Inc.

Emam, K.E., Drouin J., and Melo, W. 1998. *SPICE: The Theory and Practice of Software Process Improvement and Capability Determination*. Los Alamitos, CA: IEEE Computer Society Press.

Eriksson, H-E and Penker, M. 2000. *Business Modeling with UML: Business Patterns at Work*. New York: John Wiley & Sons Ltd.

Fowler, M. 1997. *Analysis Patterns: Reusable Object Models*. Menlo Park, CA: Addison Wesley Longman, Inc.

Fowler, M. 1999. *Refactoring: Improving the Design of Existing Code*. Menlo Park, CA: Addison Wesley Longman, Inc.

Fowler, M. and Scott, K. 1997. *UML Distilled: Applying the Standard Object Modeling Language*. Reading, MA: Addison Wesley Longman, Inc.

Gamma, E., Helm, R., Johnson, R., and Vlissides, J. 1995. *Design Patterns: Elements of Reusable Object-Oriented Software*. Reading, MA: Addison Wesley Longman, Inc.

Gilb, T. and Graham, D. 1993. *Software Inspection*. Reading, MA: Addison Wesley Longman, Inc.

Goldberg, A. and Rubin, K.S. 1995. *Succeeding With Objects: Decision Frameworks for Project Management*. Reading, MA: Addison Wesley Longman, Inc.

Grady, R.B. 1992. *Practical Software Metrics for Project Management and Process Improvement*. Englewood Cliffs, NJ: Prentice Hall, Inc.

Graham, I., Henderson-Sellers, B., and Younessi, H. 1997. *The OPEN Process Specification*. New York: ACM Press Books.

Graham, I., Henderson-Sellers, B., Simons, A., and Younessi, H. 1997. *The OPEN Toolbox of Techniques*. New York: ACM Press Books.

Hammer, M. and Champy, J. 1993. *Reengineering the Corporation: A Manifesto for Business Revolution*. New York: Harper Collins Publishers Inc.

Highsmith III, J.A. 2000. *Adaptive Software Development: A Collaborative Approach to Managing Complex Systems*. New York: Dorset House Publishing.

Hohmann, L. 1996. *Journey of the Software Professional: The Sociology of Computer Programming*. Upper Saddle River, NJ: Prentice Hall PTR.

Humphrey, W.S. 1997. *Managing Technical People: Innovation, Teamwork, and the Software Process*. Reading, MA: Addison Wesley Longman, Inc.

Jacobson, I., Booch, G., and Rumbaugh, J., 1999. *The Unified Software Development Process*. Reading, MA: Addison Wesley Longman, Inc.

Jacobson, I., Christerson, M., Jonsson, P., and Overgaard, G. 1992. *Object-Oriented Software Engineering — A Use Case Driven Approach*. New York: ACM Press.

Jacobson, I., Griss, M., and Jonsson, P. 1997. *Software Reuse: Architecture, Process, and Organization for Business Success*. New York: ACM Press.

Jones, C. 1996. *Patterns of Software Systems Failure and Success*. Boston, MA: International Thomson Computer Press.

Karolak, D.W. 1996. *Software Engineering Risk Management*. Los Alimitos, CA: IEEE Computer Society Press.

Kruchten, P. 2000. *The Rational Unified Process: An Introduction Second Edition*. Reading, MA: Addison Wesley Longman, Inc.

Larman, C. 1998. *Applying UML and Patterns: An Introduction to Object-Oriented Analysis and Design*. Upper Saddle River, NJ: Prentice Hall PTR.

Lorenz, M. and Kidd, J. 1994. *Object-Oriented Software Metrics*. Englewood Cliffs, NJ: Prentice Hall, Inc.

Maguire, S. 1994. *Debugging the Development Process*. Redmond, WA: Microsoft Press.

Marick, B. 1995. *The Craft of Software Testing: Subsystem Testing Including Object-Based and Object-Oriented Testing*. Englewood Cliff, NJ: Prentice Hall, Inc.

Mayhew, D.J. 1992. *Principles and Guidelines in Software User Interface Design*. Englewood Cliffs, NJ: Prentice Hall, Inc.

McClure, C. 1997. *Software Reuse Techniques: Adding Reuse to the Systems Development Process*. Upper Saddle River, NJ: Prentice Hall, Inc.

McConnell, S. 1996. *Rapid Development: Taming Wild Software Schedules*. Redmond, WA: Microsoft Press.

Meyer, B. 1995. *Object Success: A Manager's Guide to Object Orientation, Its Impact on the Corporation and Its Use for Engineering the Software Process*. Englewood Cliffs, NJ: Prentice Hall, Inc.

Meyer, B. 1997. *Object-Oriented Software Construction, Second Edition*. Upper Saddle River, NJ: Prentice Hall PTR.

Mowbray, T. 1997. "Architectures: The Seven Deadly Sins of OO Architecture." *Object Magazine,* April 1997, 7(1). New York: SIGS Publishing, pp. 22–24.

Page-Jones, M. 1995. *What Every Programmer Should Know About Object-Oriented Design*. New York: Dorset-House Publishing.

Page-Jones, M. 2000. *Fundamentals of Object-Oriented Design in UML*. New York: Dorset-House Publishing.

Reifer, D. J. 1997. *Practical Software Reuse: Strategies for Introducing Reuse Concepts in Your Organization*. New York: John Wiley & Sons, Inc.

Rogers, G. 1997. *Framework-Based Software Development in C++*. Englewood Cliffs, NJ: Prentice Hall, Inc.

Royce, W. 1998. *Software Project Management: A Unified Framework*. Reading, MA: Addison Wesley Longman, Inc.

Rumbaugh, J., Jacobson, I., and Booch, G. 1999. *The Unified Modeling Language Reference Manual*. Reading, MA: Addison Wesley Longman, Inc.

Siegel, S. 1996. *Object Oriented Software Testing: A Hierarchical Approach*. New York: John Wiley & Sons, Inc.

Software Engineering Institute. 1995. *The Capability Maturity Model: Guidelines for Improving the Software Process*. Reading MA: Addison Wesley Longman, Inc.

Szyperski C. 1998. *Component Software: Beyond Object-Oriented Programming*. New York: ACM Press.

Taylor, D.A. 1995. *Business Engineering with Object Technology*. New York: John Wiley & Sons, Inc.

Vermeulen, A., Ambler, S. W., Bumgardner, G., Metz, E., Misfeldt, T., Shur, J., and Thompson, P. 2000. *The Elements of Java Style*. New York: Cambridge University Press.

Warner, J. and Kleppe, A. 1999. *The Object Constraint Language: Precise Modeling with UML*. Reading, MA: Addison Wesley Longman, Inc.

Webster, B.F. 1995. *Pitfalls of Object-Oriented Development*. New York: M&T Books.

Whitaker, K. 1994. *Managing Software Maniacs: Finding, Managing, and Rewarding a Winning Development Team*. New York: John Wiley & Sons, Inc.

Whitmire, S.A. 1997. *Object-Oriented Design Measurement*. New York: John Wiley & Sons, Inc.

Whitten, N. 1995. *Managing Software Development Projects: Formula for Success, 2nd Edition*. New York: John Wiley & Sons, Inc.

Wiegers, K. 1996. *Creating a Software Engineering Culture*. New York: Dorset House Publishing.

Wiegers, K. 1999. *Software Requirements*. Redmond, WA: Microsoft Press.

Wilkinson, N.M. 1995. *Using CRC Cards: An Informal Approach to Object-Oriented Development*. New York: SIGS Books.

Wirfs-Brock, R., Wilkerson, B., and Wiener, L. 1990. *Designing Object-Oriented Software*. Englewood Cliffs, NJ: Prentice Hall, Inc.

Yourdon, E. 1997. *Death March: The Complete Software Developer's Guide to Surviving "Mission Impossible" Projects*. Upper Saddle River, NJ: Prentice Hall, Inc.

Web Resources

CETUS Links   http://www.cetus-links.org

The OPEN Website   http://www.open.org.au

The Process Patterns Resource Page
http://www.ambysoft.com/processPatternsPage.html

Rational Unified Process   http://www.rational.com/products/rup

Software Engineering Institute Home Page   http://www.sei.cmu.edu

Index

Numerics

10–20% level 65, 278
24/7 operation 8

A

abstract prototyping 101
activity modeling 37
    flow 24
Ada 170, 255
adaptor pattern 53
Albrecht, Allan J. 236
Alexander, Christopher 50
alternative
    analysis 184
    evaluation 216
Analysis and Design workflow 14
    and Requirements workflow 66
    metrics for 167
analysis error 68, 129, 152
annual reviews 243
antipatterns 158, 167

architecture 46, 170, 251
    candidate 14
    organization/enterprise-level 19
    technical 68
artifact model 24
artifacts 204
    for outsourcing 210
assumptions 68

B

Beck, Kent 25, 145
benefits
    qualitative 160
best practices 250, 272
beta testing 127
Beyer, Hugh 24, 285
Bianco, Nicole 68, 123, 285
blankfilling 169, 239
Boeing 777 170
Booch, Grady 245
brainstorming 169, 239
briefback 242
Brooks, Frederick 277

Printed and bound by CPI Group (UK) Ltd, Croydon, CR0 4YY

21/10/2024

01777097-0001